SpringerBriefs in Computer Science

SpringerBriefs present concise summaries of cutting-edge research and practical applications across a wide spectrum of fields. Featuring compact volumes of 50 to 125 pages, the series covers a range of content from professional to academic.

Typical topics might include:

- A timely report of state-of-the art analytical techniques
- A bridge between new research results, as published in journal articles, and a contextual literature review
- A snapshot of a hot or emerging topic
- An in-depth case study or clinical example
- A presentation of core concepts that students must understand in order to make independent contributions

Briefs allow authors to present their ideas and readers to absorb them with minimal time investment. Briefs will be published as part of Springer's eBook collection, with millions of users worldwide. In addition, Briefs will be available for individual print and electronic purchase. Briefs are characterized by fast, global electronic dissemination, standard publishing contracts, easy-to-use manuscript preparation and formatting guidelines, and expedited production schedules. We aim for public-cation 8–12 weeks after acceptance. Both solicited and unsolicited manuscripts are considered for publication in this series.

**Indexing: This series is indexed in Scopus, Ei-Compendex, and zbMATH **

James Neve

Reciprocal
Recommender Systems

 Springer

James Neve
Eureka Inc.
Tokyo, Japan

ISSN 2191-5768 ISSN 2191-5776 (electronic)
SpringerBriefs in Computer Science
ISBN 978-3-031-85102-5 ISBN 978-3-031-85103-2 (eBook)
https://doi.org/10.1007/978-3-031-85103-2

Preface

In the modern world, we increasingly rely on the Internet to connect with each other. Online dating services are now responsible for one third of marriages in the USA. As these services grow to tens of millions of users, we face the challenge of trying to ensure that two people who might impact each other's lives don't miss the chance to connect.

Reciprocal recommender systems help with this problem by identifying a user's preferences from their behaviour, and finding other users who would be a good fit. Most people are familiar with recommender system technology on streaming services: after watching a few movies, we are provided with suggestions for what to watch next. Reciprocal systems aim to solve the more difficult problem of predicting mutual interest between two parties who might have different levels of preference for each other.

In many ways, the stakes are much higher for reciprocal recommender system designers! The consequences of a successful movie recommendation might be someone having an enjoyable evening. Reciprocal recommendations have the potential to initiate relationships, friendships and careers. This was my primary motivation for completing a PhD on the topic, and for continuing to research the field in industry since then.

This book aims to provide an introduction to reciprocal recommendation. We start with theory, and then move on to concrete examples of the most successful algorithms in the field. Researchers and developers with a little background in machine learning should find many of the algorithms are straightforward to implement, and code samples are included to help with this.

In addition to accessible algorithms, this book also examines some more cutting edge research such as the recent interest in applying Matching Theory to reciprocal recommendation. I hope that this is interesting both to developers who are looking to optimise their systems, and to researchers who might find avenues to further advance the field and develop new methods of recommending people to people.

By the end of this book, you'll have a comprehensive understanding of the state-of-the-art in reciprocal recommendation, equipped to design and implement your own systems from scratch.

Chapter 1 provides an introduction to recommender systems and reciprocal recommendation. We look at general models of recommendation, as well as applications of reciprocal systems and challenges they present over and above standard user-item recommendation.

Chapter 2 provides some background theory on user-item recommender systems, as well as a couple of machine learning techniques which are used in subsequent chapters. It also describes aggregation functions and evaluation methods which are universal across reciprocal systems.

Chapter 3 covers collaborative filtering systems. We start here because, although they are slightly more complex than content-based filtering systems, they offer the best trade-off between strong performance metrics and simplicity of implementation, and most RRSs implemented will be collaborative filtering systems.

Chapter 4 covers content-based filtering systems. It starts with a description of the oldest reciprocal recommender system in the literature, and moves on to describe more modern systems based on unstructured information.

Chapter 5 looks at hybrid recommender systems. This includes conceptual models of combining multiple types of RRS, and some examples of how hybrid systems have been successfully used to solve specific problems.

Chapter 6 examines Matching Theory approaches to reciprocal recommendation. We start with the Stable Marriage Problem formulation by Gale and Shapley and their algorithm, and move on to more flexible formulations of the problem, which expect more sophisticated solutions.

Chapter 7 looks at some of the ethical concerns surrounding reciprocal recommendation that developers should be aware of. It also discusses future directions for reciprocal recommender systems research.

The views and opinions expressed in this book are solely my own and do not necessarily reflect the position of my employer, *Eureka Inc*.

Tokyo, Japan   *James Neve*
                                                                                              December 2024

Acknowledgements

I'd like to thank my family, Konatsu, Lily and Hugo, for supporting me during the time it took to write this book. In particular our newborn Hugo, who was happy to sleep on my lap for hours at a time while I wrote with a laptop balanced on my knees!

I'd also like to thank my friends Dr. Ercan Ezin and Yingke Shan for taking the time to read a draft of this book and for their invaluable advice.

Contents

Chapter 1
Introduction to Reciprocal Recommender Systems

1.1 Introduction

The Internet allows people to connect and interact without meeting in person. Early on, this was usually done via forums and message boards, but paradigms which encouraged one-on-one interaction such as online dating services quickly evolved. Modern iterations of these services often have millions of registered users, and therefore rely on both attribute-based searches and recommendations to help users find positive interactions with each other.

A *Recommender System* filters information to suggest items that users might find interesting. Ideally, these items are different from those that the user would have searched for themselves. We encounter recommender systems everywhere on the Internet, but the most common examples are shopping services such as *Amazon*, which suggest items to us that we might want to buy based on our purchase history and ratings.

For example, if I buy baby clothes on a shopping service, a simple algorithm might recommend me more baby clothes. There are any number of ways of making this system more useful, however. A slightly more sophisticated system would be able to infer from other users' behaviour that if I buy baby clothes, I might also need nappies, formula milk and so on. If the system is looking at behaviour over time, I might even find myself recommended age-appropriate clothes and toys six months later.

Person-to-person recommender systems, or *Reciprocal Recommender Systems* [9] (RRS), are distinct from the item recommender systems used by services such as *Amazon* and *Netflix*. User-item recommender systems make recommendations considering the user's preferences and the item's attributes, whereas RRSs must consider both sides of the preference equation.

RRSs start from the same basic principle as user-item recommender systems. The most reliable way to find out a user's preferences is their preference history. A user who likes other users in certain age ranges, or with certain hobbies, is likely to respond positively to recommended users with similar attributes. Users might

J. Neve, *Reciprocal Recommender Systems*, SpringerBriefs in Computer Science, https://doi.org/10.1007/978-3-031-85103-2_1

make this information *explicit*. For example, on an online dating service, Alice might write on her profile that she's looking for users in the 20 to 30 age range. Alternatively, we may infer this from *implicit* cues. If Alice does not write anything on her profile, but exclusively sends messages to users under 30, we can assume that she might respond positively to recommendations in this range. After establishing unidirectional preferences, systems attempt to balance both sides of the preference equation by recommending a user who is likely to be reciprocally interested.

Preference data in RRSs is usually implicit. Users do not give each other star ratings as they might to movies on a streaming service, but interact directly. An RRS aims to encourage positive interactions while minimising negative ones. It is up to the system designer to decide how interactions are counted as positive or negative, and this will ideally depend on the environment: a casual conversation that ends without a face-to-face meeting might be considered successful by some systems and some users, and unsuccessful by others.

Some challenges in this domain aren't immediately intuitive. For instance, there is no problem with recommending a popular item on a shopping service. However, popular users on online dating services may receive thousands of messages and only respond to a handful, so recommending them may not result in a productive reciprocal interaction [4].

While there is extensive research on the design and development of recommender systems, RRSs are significantly sparser in the literature [7]. Much of this is due to data sharing ethics: datasets of anonymous product or movie preferences are easy to release, but data from user interaction-based services is much more sensitive, and cannot usually be released anonymously.

This book will describe, with examples, the design of reciprocal recommender systems, based on the most up-to-date research. We will discuss some theoretical background, and then different paradigms including collaborative, content-based and hybrid systems, with examples of those that have been successful in the real world. This chapter provides an overview of user-item recommender systems and RRSs, including applications, design paradigms and specific challenges not faced by user-item recommender systems.

This book does not require a deep knowledge of user-item recommender systems, but it does assume a basis in Computer Science theory, including a knowledge of formal logic notation. Code examples in this book are in *Python*, which is a language most machine learning developers are familiar with, but examples do not rely on any complex facets of the *Python* language, and should be readable by anyone with coding experience.

1.2 Applications of Reciprocal Recommender Systems

1.2.1 Online Dating

The earliest and broadest applications of RRSs have been in online dating [8]. For online dating, matching users is the primary objective of the service so there is high motivation to develop specialised algorithms which can effectively do this. Dating services generally possess rich datasets. A standard user profile on an online dating service contains structured data such as age, profession and location, in addition to unstructured data such as a freetext profile and photograph. RRSs can take advantage of all of this information in making recommendations. It is also interesting to analyse existing RRSs to see which elements of user profiles on online dating services are most valuable in guiding how users make decisions about each other.

In addition, modern dating services tend to have a clear process by which users interact with each other, from which we can extract implicit indicators of positive or negative preference. It is common to require two users to send preference indicators to each other (usually called *Likes*) and *Match* before they can communicate with each other. In RRSs, the Match rate is the primary performance indicator. Implicit negative preference indicators often also exist on online dating services. A popular example of this is *Tinder*, where users swipe right to accept and left to reject the user that is presented.

Finally, online dating services tend to have very large quantities of data. Popular services have upwards of tens of millions of registered users. This allows both for representative training and analysis of results, and for recommenders to effectively be built based on subsets of users (for instance, specific RRSs for users living in certain age ranges and locations).

The quantity and quality of data therefore make online dating services rich targets for RRS development. Due to valid privacy concerns these datasets are very rarely made public, but where collaborations between researchers and companies who hold them have been possible, recommender systems with a high level of accuracy have been developed.

1.2.2 Recruitment

Reciprocal recommendation for recruitment is based on matching companies and job openings with potential applicants for them. Recruitment provides a slightly different challenge from online dating, as two conceptually different entities are being recommended to each other: an applicant and a company. Perhaps because of this there are relatively fewer examples of job recommender systems in the literature. Nonetheless, there are interesting differences, which makes it a worthy field of study. Companies often specify clearly what they are looking for from applicants, and applicants often write their own qualifications in a standardized format. Additionally, there is less reliance than online dating on unstructured data such as photos.

Until recently, applications of RRSs to recruitment services was in the minority. However, the recent advances in applying Matching Theory to RRSs has spurred renewed interest. In spite of the romantic context of the Stable Marriage Problem, the formulation more closely resembles a recruitment service than an online dating service. A Match between a job posting and a candidate often removes both the candidate and the job posting from the pool. Users in online dating contexts prefer to Match with as many other people as possible, so finding a single optimal set of pairings is less valuable.

1.2.3 Social Applications

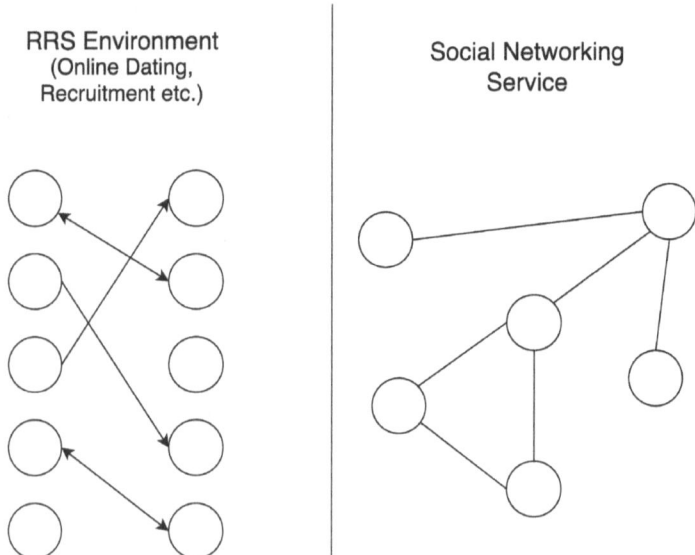

Fig. 1.1: How RRSs and social networking services view users

There are a number of other fields in which reciprocal recommendation is applied. Social network websites are, for some definition, applications of reciprocal recommender systems, and in particular social networks which encourage one-on-one meetings such as conversation exchange services have applied RRSs. However, many large scale social networking services such as Facebook view relationships as a graph and introduce users to each other based on the number and weight of shared connections, under the assumption that they already know each other. Figure 1.1 illustrates this difference: in short, RRS environments view user relationships as a directed, bipartite graph, whereas social networks are neither bipartite nor directed. The algorithms used by social networking services are therefore significantly different, and are outside the scope of this book.

A final field where significant RRS research has been carried out is online learning. Online learning platforms often aim to form one-on-one connections between people, either between two learners who are at a similar stage and might benefit from each other's experience, or between a teacher and a student. RRSs are more applicable to this problem.

1.3 Models of Reciprocal Recommender Systems

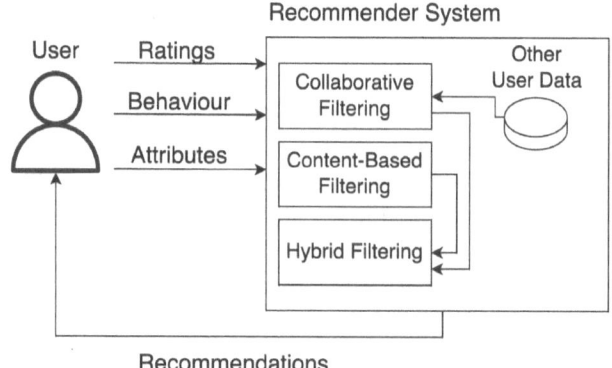

Fig. 1.2: General conceptual model for Recommender Systems

As shown in Figure 1.2, recommender systems take as input various information about users including their previous ratings, behaviour and attributes. Using this information, for a given user Alice, they create a ranked list of items on the service. The items at the top of the list are shown to her as recommendations. To create this list, we compute a *Preference Score* for each item. This is a score between 0 and 1 that represents how much we expect Alice to like that item. A higher score represents a better recommendation. The field of Recommender Systems looks for the most accurate way to compute preference scores.

There are a number of paradigms of recommender systems, but the three most commonly used and successful ones are *Collaborative Filtering*, *Content-Based Filtering* and *Hybrid* systems [1]. Collaborative filtering scores potential recommendations based on correlations between users. Content-based filtering computes scores based on similarity between the attributes of a potential recommendation, and items that the user has liked in the past. Hybrid systems blend different algorithms to take advantage of their individual strengths. Each of these is given a more detailed overview in this section, and has a chapter dedicated to it later in this book.

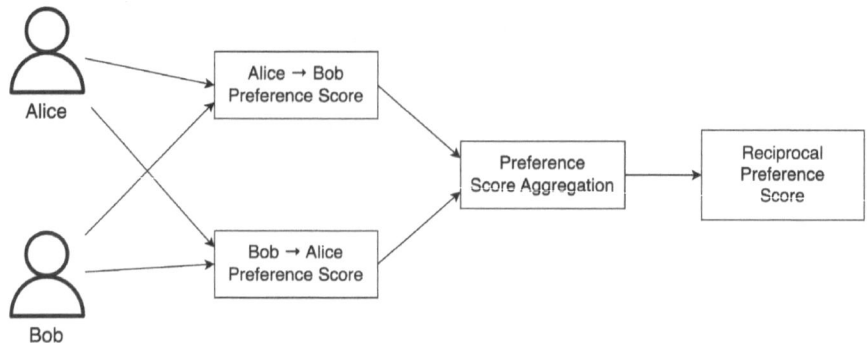

Fig. 1.3: Reciprocal Recommender Systems Outline

The main difference between RRSs and user-item recommender systems is the bidirectional nature of the former. RRSs have to take account of both sides of the preference equation, in order to make recommendations that will be reciprocated. Knowing Alice's preference for Bob is only useful if we know whether Bob also likes Alice. The standard design methodology is to adapt user-item recommender system principles to calculate two preference scores, one for Alice's preference for Bob and one for Bob's preference for Alice. The two preference scores are then aggregated into a single reciprocal preference score, which represents the users' preference towards each other. This reciprocal preference score, calculated for pairs of users across the service in which it is being deployed, is then used to make recommendations. This process is shown in Figure 1.3.

RRS design therefore generally consists of two key choices: the recommender system to determine unidirectional preferences, and the aggregation function used to combine the unidirectional preference scores into reciprocal preference scores [6]. This section provides an overview of the models used to calculate preference scores, which will be familiar to readers experienced in user-item recommender system design.

Approaches based on *Matching Theory* have been increasingly prominent in recent years. Instead of basing their recommendations on preference scores, these algorithms make recommendations by aiming to find optimal pairings of users. They are also introduced, following the sections on more traditional methods.

1.3.1 Collaborative Filtering

Collaborative filtering uses correlations between user preferences to make recommendations. For example, on a given service, Alice likes users Bob, Charlie and David. Kate likes Bob and Charlie. Based on their shared preference for Bob and Charlie, we can infer that Alice and Kate have similar preferences. We might therefore recommend David to Kate.

Finding similar users is usually done by calculating *Similarity Scores* between pairs of users. A similarity score is a number which represents how similar two users are based on their historical preferences. We can use these scores to find the most similar users to Alice, and her preference scores will then be computed from the preferences of those users.

Although this is easy to do with small datasets and produces effective recommendations, it becomes more complex as the size of the dataset increases, and the number of computations required to compute similarity scores across all pairs of users and items becomes infeasible. Increasingly effective strategies for solving this problem, and for similarities from large datasets, have been a key part of the advancement of recommender systems over the last decade.

Experimental results have shown that in user-item domains, collaborative filtering is the most effective type of recommender system, once sufficient data is available. This is also generally true in reciprocal domains.

Collaborative filtering does struggle with the *Cold Start Problem*, where users with little data are not given effective recommendations (and in reciprocal domains, not included in other users' recommendation lists). This is because a small number of preference expressions is usually not enough to establish clear enough similarities with other users to make effective recommendations. This problem is particularly significant in the case of online dating, where users are extremely likely to give up on a service quickly if they do not perceive that they can find matches there.

1.3.2 Content-Based Filtering

In content-based filtering, the attributes of users and items are used to make recommendations. User preference for other users is viewed in terms of preference for their individual attributes. For example, if Alice has shown preference for five users, and four of them kept dogs as pets, a content-based RRS might infer that Alice likes people who have dogs, and recommend her users with this attribute. Content-based filtering uses a user's preference history to build a model for that user's preference for individual attributes, and use this to predict future preferences.

In some cases, these models are relatively simple. Early RRSs used only categorical data such as age and location to build preference models, for which purpose traditional neighbourhood models of content-based filtering can easily be adapted from user-item recommender systems. More recent systems have attempted to use unstructured data such as user photos to make predictions, which involves more complex models.

Content-based models have the advantage of often being simpler to implement and scale than other methods of recommendation. They can also be effective at dealing with Cold Start situations: basing recommendations on attributes requires only that a user fill in their profile to start appearing in recommendation lists. This is often particularly useful in online dating scenarios: a user appearing in recommendation lists and receiving *Likes* immediately tends to increase the likelihood that they will stay on the service and interact with other users. This is less likely to happen with collaborative filtering because of the Cold Start problem.

However, content-based methods do have disadvantages, particularly in RRS scenarios. They tend to be less effective at accurately predicting preferences than collaborative filtering methods. Services almost always provide attribute-based search to users, and users are generally aware of their own preferences. In addition, while recommenders based on categorical data are easy to design, research has shown that in online dating services, categorical data is usually not a large indicator of preference; photos and free text profiles tend to have a much larger degree of importance. Designing a recommender based on these attributes tends to be much more complex.

1.3.3 Hybrid Filtering

Hybrid RRSs combine two or more collaborative filtering and content-based filtering recommender systems to leverage the advantages of both methods [2]. There are a number of different ways of combining systems. For example, *Weighted* systems decide a reciprocal preference score by taking a weighted average of the scores from multiple different recommender systems. *Switching* systems decide which score to use depending on the context. For example, we can try to use this to tackle the Cold Start problem by using, for example, a content-based system for initial recommendations, and then using collaborative filtering once the user has a more informative preference history.

In the world of user-item recommender systems, hybrid systems tend to be the most effective in terms of performance on offline metrics. However, they are also the most complex to design, and rely on effectively combining existing successful systems in such a way as to overcome the weaknesses of individual systems. In the case of reciprocal recommendation, where there is not as much information about which algorithms are the most effective, and work is often done on various private datasets which makes it difficult to compare or reproduce work, hybrid systems become even more challenging to effectively design.

1.3.4 Matching Theory Methods

Matching Theory is a field which originated with a problem proposed by Gale and Shapley in 1962: the *Stable Marriage Problem* [3]. If we have two sets of men and women of equal sizes, and each of them has submitted a list ranking potential partners in the opposite list in order of preference, how do we optimally match men and women such that there is no man and woman who are not paired with each other, and who would prefer each other to their assigned partners.

The problem's application to reciprocal recommendation, and its limitations are immediately apparent. Solutions to the problem aim to maximise reciprocal preference, which is also the goal of RRSs. But the formulation of the problem in a strict sense where every user ranks every other user is hard to draw analogies with online dating or recruitment, where users usually send binary indicators of preference to small numbers of other users and do not provide convenient preference lists. The field that evolved from this problem is Matching Theory, which contains formulations that are more flexible and more widely applicable.

Matching Theory-based solutions to reciprocal recommendation have recently been attracting attention, and appear to be quite successful [10]. It is still unclear whether they outperform standard collaborative filtering methods (or whether the performance gain is worth the additional complexity) when implemented into real systems, but nonetheless they represent an interesting alternative approach, distinct from adapting user-item methods.

1.4 Challenges for Reciprocal Recommender Systems

1.4.1 Data Structure

An effective recommender system predicts user preferences; in the case of an RRS it predicts the preferences of users for each other. In user-item recommender systems, users often explicitly express their preference for items by *rating* them. *Amazon*, for example, provides a star rating scale for users to indicate how strong their preference for or against an item is. *Netflix* also lets users explicitly express their positive or negative preference for a series or movie. Online dating services, on the other hand, almost never provide a method for users to explicitly rate each other. We therefore have to design recommender systems based on implicit indicators of preference.

Many online dating services allow users to *Like* each other as an initial interaction. An exchange of Likes leads to a *Match*, which subsequently allows users to exchange messages with each other. A Like may constitute a positive indicator of preference; a user seeing and ignoring another user's Like may be considered a negative preference indicator. We usually call this a *Nope*. Although this formula is not universal, enough modern online dating services use this or some close variation that we can usefully adopt this terminology for the sake of formalising algorithms. For example, on a service where users do not send Likes but instead are given the option of sending messages directly, we can consider an initial message from one user to another to be a Like, and a positive response to be a Match.

It is not clear, however, how far using Likes and Matches as key performance indicators of a recommender system might lead to users achieving the ultimate goal of using the service. Where the end goal of a user-item recommender system might be to help a user buy a product they like, or find a movie to watch, the end goal of a user matching service might be to help users find a long-term relationship or marriage. More research is required to determine the extent of the correlation between increasing the number of Matches, and increasing the number of meaningful relationships.

1.4.2 Fairness

In user-item recommender systems, popular items do not present a significant problem to the system. Popular items are often a good thing to recommend, because they have a track record of making a large number of users happy. In reciprocal environments, the opposite is true. Popular users on social services, and especially in online dating services, often receive far more communication than they can realistically respond to, and the the probability of an ordinary user's indicator of preference being reciprocated by a popular user might be as little as one in a thousand. This is particularly a problem for collaborative filtering because popular users tend to commonly appear in preference histories, and because of this they have a high chance to be recommended. Conversely, users who have received very few indicators of preference (including new users) are very unlikely to be recommended, and this becomes a self-reinforcing situation which may eventually cause these users to leave the service.

Because of this, RRS environments have to consider *fairness* when they make recommendations. This usually means having a system in place which reduces the chance of popular users being recommended and increases the chance of less popular users being recommended. This is a difficult balance to strike, however: adjusting recommendations too far in favour of fairness may come at the expense of recommendatinos which reflect the target user's true preferences.

1.4.3 Preference Aggregation

As outlined in Section 1.3, for two users Alice and Bob, an RRS will usually generate two preference scores, representing Alice's estimated preference for Bob and Bob's estimated preference for Alice. In order to make effective reciprocal recommendations, these two scores need to be combined into a single reciprocal preference score which can be used to rank recommendations. This is done by an *aggregation operator*. In reciprocal systems, the choice of aggregation operator influences predictive power [5].

Many RRSs use simple averages as the method of aggregating preferences between two users. However, more sophisticated methods have been tested, and the aggregation operator used does make a difference to the performance of the system. In addition, the choice of operator is not clear cut, and appears to depend on the system being used.

References

1. Aggarwal, C.C., et al.: Recommender systems, vol. 1. Springer (2016)
2. Burke, R.: Hybrid recommender systems: Survey and experiments. User modeling and user-adapted interaction **12**, 331–370 (2002)
3. Gale, D., Shapley, L.S.: College admissions and the stability of marriage. The American Mathematical Monthly **69**(1), 9–15 (1962)
4. Kleinerman, A., Rosenfeld, A., Ricci, F., Kraus, S.: Optimally balancing receiver and recommended users' importance in reciprocal recommender systems. In: Proceedings of the 12th ACM Conference on Recommender Systems, pp. 131–139 (2018)
5. Neve, J., Palomares, I.: Aggregation strategies in user-to-user reciprocal recommender systems. In: 2019 IEEE International Conference on Systems, Man and Cybernetics (SMC), pp. 4031–4036. IEEE (2019)
6. Neve, J., Palomares, I.: Latent factor models and aggregation operators for collaborative filtering in reciprocal recommender systems. In: Proceedings of the 13th ACM conference on recommender systems, pp. 219–227 (2019)
7. Palomares, I., Neve, J., Porcel, C., Pizzato, L., Guy, I., Herrera-Viedma, E.: Reciprocal recommender systems: Analysis of state-of-art literature, challenges and opportunities towards social recommendation. Information Fusion **69**, 103–127 (2021)
8. Pizzato, L., Rej, T., Akehurst, J., Koprinska, I., Yacef, K., Kay, J.: Recommending people to people: the nature of reciprocal recommenders with a case study in online dating. User Modeling and User-Adapted Interaction **23**, 447–488 (2013)
9. Pizzato, L., Rej, T., Chung, T., Koprinska, I., Kay, J.: Recon: a reciprocal recommender for online dating. In: Proceedings of the fourth ACM conference on Recommender systems, pp. 207–214 (2010)
10. Su, Y., Bayoumi, M., Joachims, T.: Optimizing rankings for recommendation in matching markets. In: Proceedings of the ACM Web Conference 2022, pp. 328–338 (2022)

Chapter 2
Theoretical Background

2.1 Introduction

This chapter outlines the background theory required for reciprocal recommender systems. As outlined in the previous chapter, RRSs are often designed based on conventional user-item recommender systems. Where specific algorithms are explained in subsequent chapters, this is done from first principles, so a deep understanding of recommender systems in general is not required. Understanding the basics of how user-item recommender systems are constructed does, however, make this process a little easier. Much of this chapter is dedicated to the basics of user-item recommender systems, and readers already comfortable with this can skip to Section 2.4, which is the start of the material unique to reciprocal recommendation.

We start by looking at collaborative filtering (recommendation based on similar users). This chapter gives background on *Nearest Neighbour* collaborative filtering, which is one of the earlier forms. We then move on to content-based filtering (recommendation based on user preference for attributes). For both, we explain simple methods of learning user preferences using easy-to-calculate metrics. In the case of content-based filtering, we also outline some of the background for how user profile data can be converted into machine-readable information using machine learning methods.

Preference aggregation (the process of combining two unidirectional preference scores into a single bidirectional preference score) is also an important part of reciprocal recommendation. Combining the two preferences is done by an aggregation function, and the choice of function impacts the effectiveness of the recommender system [9]. We therefore also examine some commonly used aggregation functions.

Finally, RRSs are evaluated differently from user-item recommender systems. Unidirectional indicators of preference are not a reliable indicator of how well the system is working, when bidirectional preference relations are the objective. This chapter ends with a discussion of evaluation functions for RRSs.

© The Author(s), under exclusive license to Springer Nature Switzerland AG 2025
J. Neve, *Reciprocal Recommender Systems*, SpringerBriefs in Computer Science,
https://doi.org/10.1007/978-3-031-85103-2_2

2.2 Collaborative Filtering

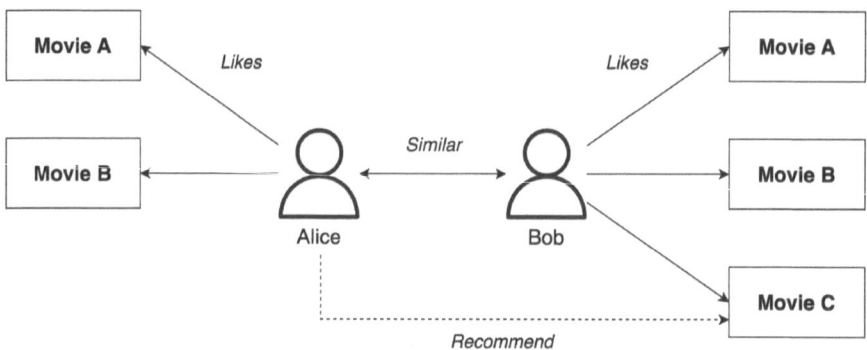

Fig. 2.1: Collaborative Filtering

Collaborative filtering uses correlations between users to make recommendations. This concept is illustrated in Figure 2.1: if Alice likes movies A, B and C, and Bob likes movies A and B we might recommend him C based on his similar preferences to Alice. Users often like items for reasons beyond the properties listed on the service. For example, the movies might be different genres, but the same actor might appear in all three of them. Collaborative filtering is able to capture this, even if the information about the actors in a movie is not listed on the service.

The earliest forms of collaborative filtering were *neighbourhood Models* (otherwise known as *kNN*) [1]. These models compute similarity between pairs of users. When we want to predict Alice's rating for the movie *The Matrix*, we consider how her k most similar users have rated it. If users with similar taste in movies (i.e. similar preference histories) liked it, our hypothesis is that Alice might also like it, and we can display it to her as a recommendation.

For m users and n items, we consider an incomplete $m \times n$ matrix R of ratings. Then for recommendation, we want to predict r_{uj}, the missing rating of user u for item j. neighbourhood models do this prediction in two steps:

1. A *Similarity Metric* is calculated between user u and other users on the system, based on the preferences those users have expressed for items so far. This is a number between -1 and 1, where -1 indicates a perfect negative correlation (i.e. that the target user's ratings are the opposite of u) and 1 indicates that the two users have exactly the same preferences. This similarity metric is used to identify the *top-k* users similar to u.
2. The *top-k* users' ratings of item j are aggregated into a preference score of user u for item j. This preference score, and the scores of other items that the user has not yet expressed a preference for, are used to decide which items to recommend to u.

It is worth being aware that the above describes an *user-based neighbourhood model*, where similarities between users are used to make recommendations, and that *item-based neighbourhood models* also exist, which use similarities between items to make recommendations. These will be omitted from this book, as they are not relevant to reciprocal systems, where there are no items and only users.

The first step is to define a formula which calculates the similarity metric between two users. Formally, for a service with n items, user A with ratings $(a_1...a_n)$ and user B with ratings $(b_1...b_n)$, we want a metric $S(A, B)$ which represents the similarity between user A and user B. In order to do this, we first normalise the ratings of A and B to account for the fact that some users rate things higher in general than others. For A, we calculate the arithmetic mean of their ratings $\bar{A}$:

$$\bar{A} = \frac{\sum_{i=1}^{n} a_i}{n} \tag{2.1}$$

Then the normalized rating a_i of an item i is defined as $(a_i - \bar{A})$.

There are a number of methods of calculating similarity between two users. A common method is the *Pearson Correlation Coefficient* (PCC) [7], as shown in Equation 2.2. This function is a measure of the strength and direction of a linear relationship between two variables:

$$S(A, B) = \frac{\sum_{i=1}^{n} (a_i - \bar{A})(b_i - \bar{B})}{\sqrt{\sum_{i=1}^{n} (a_i - \bar{A})^2 \sum_{i=1}^{n} (b_i - \bar{B})^2}} \tag{2.2}$$

The result of this equation is a number between -1 and 1 which represents the similarity between the ratings of the two users A and B. A higher number represents a greater similarity; 1 occurs when the users have the same ratings, while -1 occurs with exactly opposite ratings.

Movie → ——— User ↓	The Matrix	Pulp Fiction	Die Hard	Notting Hill	Love Actually
Alice	1	1	1	0	0
Bob	1	0	1	0	0
Charlie	0	1	0	1	1
David	1	1	1	?	0
Ed	1	0	0	0	1
Fred	1	1	1	1	0

Table 2.1: Ratings of users for movies

For example, see Table 2.1. This is a ratings matrix for a movie website for six users and five popular movies, with the users giving a binary rating representing positive or negative preference (imagine a 'Thumbs Up' or 'Thumbs Down' system).

We have one missing rating: David's rating for *Notting Hill*, which we'd like to estimate. We first use the PCC to calculate similarity between every pair of users. Note that in David's case, we would skip *Notting Hill* in our PCC calculation with other users.

User → User ↓	Alice	Bob	Charlie	David	Ed	Fred
Alice	1	0.667	−0.667	1.000	−0.167	0.612
Bob	0.667	1	−1.000	0.577	0.167	0.408
Charlie	−0.667	−1.000	1	−0.577	−0.167	−0.408
David	1.000	0.577	−0.577	1	−0.577	1.000
Ed	−0.167	0.167	−0.167	−0.577	1	−0.612
Fred	0.612	0.408	−0.408	1.000	−0.612	1

Table 2.2: Pearson correlation coefficients between users

The similarity scores for each user are displayed in Table 2.2. (Note that the table is symmetrical across the diagonal, but here we display all the values, as this improves readability when searching for similar users.) Having calculated similarity scores, we can then use these to make predictions about the unknown rating of a user A. We first take the k most similar users to A. Their similarity scores are then used to compute a *weighted average* of their ratings for a particular item, where the weights are their similarities to A. This forms A's preference score for that item.

For example, consider predicting the rating of David for the movie *Notting Hill*, where k is 3. We take the three most similar users to David by score. From Table 2.2, this is Alice (1.000), Fred (1.000) and Bob (0.577). David's preference score for Notting Hill, $P_{\text{David} \rightarrow \text{Notting Hill}}$ is then a weighted average of the scores of his most similar users, with their similarities as weights. That is to say, the sum of the products of the ratings with the weights, divided by the sum of the weights:

$$P_{\text{David} \rightarrow \text{Notting Hill}} = \frac{(0 \times 1.000) + (1 \times 1.000) + (0 \times 0.577)}{1.000 + 1.000 + 0.577} \qquad (2.3)$$

David's preference score for *Notting Hill* is therefore 0.39. Interested readers are encouraged to eliminate other ratings from Table 2.1 and try to predict them with similarity calculations followed by weighted averages.

Having predicted scores for unknown items, we can then use these to rank recommendations, recommending items with the highest predicted scores to users. kNN recommenders have been used in live services since the nineties, and produce effective results.

Our toy example had just a few users and movies, and a mostly complete matrix. While this is useful for demonstration purposes, it is not representative of a live service, which may have millions of users and items, and a sparse matrix (most users only rate a few items). It is relatively easy to see how time complexity might become a problem as the number of users and items increases. We need to calculate similarity between all pairs of users across all items to make any recommendations. Chapter 3 discusses this problem and potential solutions to it in reciprocal environments in more detail.

Finally, we note that it is easy to confuse preference scores, which are numbers in the interval $[0, 1]$ with probabilities. They are, however, very different concepts: preference scores are used for ranking, and do not represent the probability that a user will like an item or another user. They are useful relative to other preference scores within a recommendation context, but a single preference score outside of a ranking, however close it is to 0 or 1, does not necessarily give us very much information about user preference in absolute terms.

2.3 Content-Based Filtering

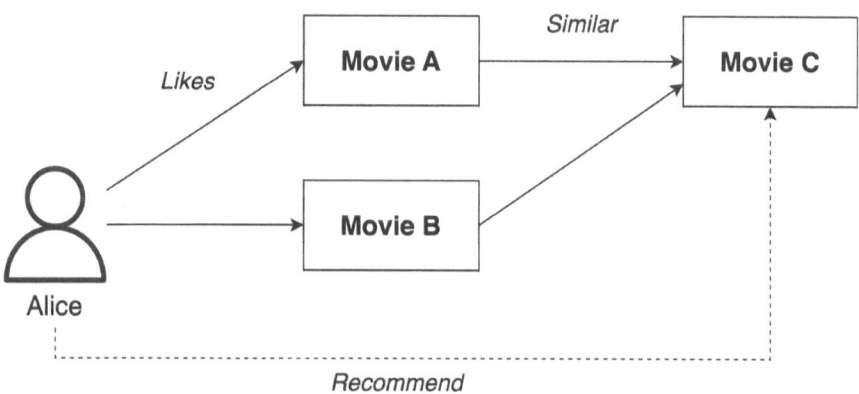

Fig. 2.2: Content-Based Filtering

Content-based filtering recommender systems learn user preferences for item attributes, and use these to make recommendations. For example, in the case of a user who joined a movie streaming service and watched several action movies from the 1980s, the recommender system might learn the user's preference for the genre and the era and recommend more of the same type of movies. This is shown in Figure 2.2. Early recommender systems were often content based systems [12], and early reciprocal systems also followed this trend. This section gives a brief introduction to content-based filtering in the context of user-item recommender systems, which can then be extended to reciprocal environments in Chapter 4.

Content-based systems make recommendations based on two types of data:

1. The *attributes* of the items in the system. This might be categorical data such as the genre of a movie, or it might be unstructured data such as a freetext description or an image.
2. The *preference profile* of the user. This describes the preferences of the user, based on their actions so far while using the service. The preference profile is generated by explicit preference expressions, such as rating a movie on a scale, or by implicit expressions, such as by watching a movie all the way through. The preference profile is subdivided into *interests*, which are numbers that represent the preference of a user for the attribute of an item.

2.3.1 Feature Extraction

The first stage of content-based recommendation is feature extraction. This is the process of turning item attributes, which often comprise structured and unstructured data, into a *vector space representation*: a list of numbers which represents the item in the context of the recommender system. In the case of structured data, such as a movie's genre or its release year, this is a simple process: discrete variables are converted to integers, continuous variables can be dealt with in a variety of ways such as bucketing.

In the case of unstructured data such as text descriptions and photos, this is more complicated. There are a number of techniques which are used to perform feature extraction on unstructured data, and Chapter 4 goes into more detail on this in the context of RRSs.

Many modern systems use machine learning methods to derive a vector representation of unstructured data, and the theory behind these methods is worth expanding on here, because it is broadly used across all types of RRS. *Convolutional Neural Networks* (CNNs) [3] are particularly adept at extracting meaning from text and images, and convert them to a meaningful vector space representation. We expand briefly on CNNs here specifically because they are a widely used method of extracting features from images, and have proven to be particularly effective. However, we note that other machine learning models such as *Support Vector Machines* [5] and *Transformers* [8] are also effective depending on the situation, and it is worth being aware of the options and trying multiple methods when searching for an effective model to create feature representations.

Because unstructured data forms such a large part of how users made decisions on especially online dating services, and because CNNs are an essential part of some of the RRS algorithms for content-based filtering, we explain them briefly here. The caveat is that a more thorough understanding of them is likely to be required to implement the algorithms in question. Readers who are not already familiar with them would benefit from reading separately about the topic, specifically for the benefit of understanding algorithms described in Chapter 4 (they are not used in other chapters).

2.3.1.1 Convolutional Neural Networks

Neural Networks are a type of machine learning model. They consist of layers of interconnected *Neurons*, which take an input, perform a mathematical transformation, and pass the result to other neurons. Traditional neural networks consist of the following:

- **Input Layer** which accepts a vector of numbers.
- **Hidden Layers** each of which accepts the output of the previous layer of Neurons, performs its transformation, and passes its output on to the next layer.
- **Output layer** which produces the final result, which might be a class or a value.

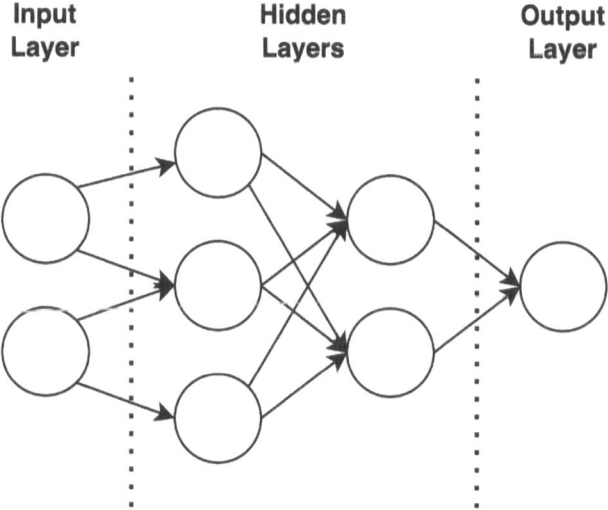

Fig. 2.3: A fully-connected Neural Network

Figure 2.3 shows the structure of a simple Neural Network with two inputs, two hidden layers and one output. It is easy to imagine that we might be able to use this to solve a simple classification or regression problem. As an example from the world of online dating, we might be able to use this to approximately predict a user's popularity on a scale of 0 to 1 given their age and their height.

Neural Networks are generally trained by *Backpropagation* [6], whereby the parameters of the mathematical transformation performed by the nodes are adjusted based on error from examples in a training set, starting with the nodes nearest the output layer, and propagating the adjustments backwards through the network to move closer in incremental steps to a network which produces the correct result.

In fully-connected Neural Networks, where every node is connected to every other node, the number of connections grows exponentially with the number of nodes. The training process becomes computationally infeasible, especially for processing data with a very large number of inputs where each color channel of each pixel often has to be a separate input, and where a sufficient number of hidden layers are required to make complex classifications such as identifying arbitrary objects.

CNNs are designed to process data with a grid structure, such as images. They generally contain three types of layers, which reduces complexity compared to fully connected structures:

- **Convolution Layers** apply filters across rectangular regions of data. These layers are used to detect patterns such as edges and textures.
- **Pooling layers** downsample the input, reducing the information in a cluster of neurons to a smaller number, which reduces complexity.
- **Fully-Connected Layers** generally occur at the end of the network, and use the extracted and downsampled features to make predictions.

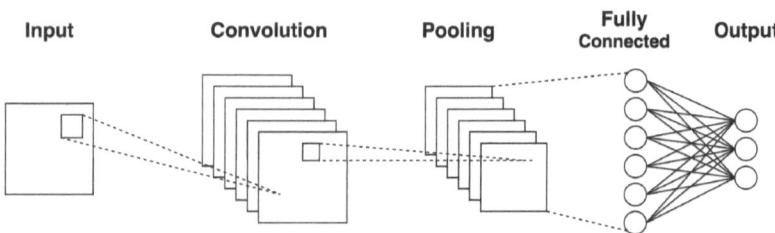

Fig. 2.4: A Convolutional Neural Network

A visual representation of a CNN is displayed in Figure 2.4. CNNs are particularly prominent in image recognition tasks, but have also been used to capture patterns in text and in time-series data such as user behaviour. This makes them useful in the context of especially content-based reciprocal recommendation, where much of the information about initial appeal of users to each other (e.g. physical attributes) is contained in unstructured data such as photos and text.

2.3.2 Learning User Profiles

After features have been extracted from item profiles and converted to vector space representations, user preferences for the features can be learned. We only cover this briefly here, as reciprocal recommendation methods for content-based filtering differ enough from user-item content-based filtering that an in-depth coverage does not provide a particularly useful basis for reciprocal recommendation.

User preferences for individual features can be used to estimate how much the user might like the item. In user-item recommender systems, this can be a classification problem based on binary data, such as whether or not a user will buy an item, or it might be a regression problem based on estimating a user's preference for an item on a scale. In RRS contexts, it is almost always a classification problem based on implicit data - services where reciprocal systems are used rarely ask users to rate each other on a scale - so examples in this chapter will focus on this.

A simple classification technique for content-based filtering is *Nearest neighbour Classification*. Nearest neighbour classification uses a *Similarity Function*, which calculates a number representing the similarity between two items based on their vector space feature representations. Items with high similarity to those that the user has already shown preference for can then be recommended. A common example of a similarity function is the *Cosine Similarity*. For two feature vectors $\overline{X} = (x_1...x_d)$ and $\overline{Y} = (y_1...y_d)$ representing the attributes of items, the cosine similarity is defined as:

$$\text{Cosine}(\overline{X}, \overline{Y}) = \frac{\sum_{i=1}^{d} x_i y_i}{\sqrt{\sum_{i=1}^{d} x_i^2}\sqrt{\sum_{i=1}^{d} y_i^2}} \tag{2.4}$$

A simple method of using this to make recommendations for Alice is to calculate the average similarity between a candidate item Y and the items which Alice has liked in the past $X_1, X_2, \ldots, X_n$. This gives us the preference of Alice A for item Y, $P_{A \to Y}$:

$$P_{A \to Y} = \sum_{i=1}^{n} \frac{\text{Cosine}(\overline{X_i}, \overline{Y})}{n} \tag{2.5}$$

We can then use this to make recommendations for Alice by ranking items in order of preference score.

2.4 Preference Aggregation

Commonly, a unidirectional recommender system is used to generate two unidirectional preference scores between two users, Alice and Bob, in the range $[0, 1]$. In order to determine whether Bob should be recommended to Alice, we use an *aggregation operator* to combine the two values into a single value which represents the strength of the bidirectional preference relation. An intuitive way of combining the two values, x_1 and x_2, is the *arithmetic mean*:

$$M(x_1, x_2) = \frac{x_1 + x_2}{2} \tag{2.6}$$

This is sometimes used in reciprocal recommendation, but has the weakness that a single extreme value can raise the score significantly. This is not usually desirable: intuitively, Alice and Bob are more likely to match if their respective preference scores are both 0.6 than if one is 0.3 and the other is 0.9, because rejection from one user is considered an overall negative result. The *Harmonic Mean* is therefore more commonly used [10, 13], which for x_1 and x_2 is defined as:

$$H(x_1, x_2) = \frac{2x_1 x_2}{x_1 + x_2} \tag{2.7}$$

This penalises difference between the two terms, so $H(0.6, 0.6)$ is 0.6, but $H(0.3, 0.9)$ is 0.45, and evidence suggests that this leads to more effective RRSs. The third of the three *Pythagorean Means* is the *Geometric Mean*, which for x_1 and x_2 is defined as:

$$G(x_1, x_2) = \sqrt{x_1 x_2} \tag{2.8}$$

This also penalises difference between inputs, but not to the extent of the harmonic mean. For example, $G(0.3, 0.9)$ is 0.518. This is presented here for the sake of completeness, but less commonly used. Finally, we present the weighted average. For a weight α this is:

$$WA(x_1, x_2) = \alpha x_1 + (1 - \alpha)x_2 \tag{2.9}$$

Reciprocal recommender environments often represent unbalanced relationships, where one side's preference is more highly valued. For example, an extremely prestigious company is almost always in the driving seat when hiring graduate applicants. A weighted average can represent this by applying a higher weight to the party whose preference plays a bigger part in determining Matches. The complexity in this case is that the parameter α must also be computed in such a way that improves the results of the system over using a simpler method of aggregation.

2.5 RRS Evaluation

There are two methods of evaluating recommender systems and RRSs: *Online* and *Offline*. In online evaluation, users are presented with recommendations generated by the system while using the service, and their reactions to those recommendations are used for evaluation purposes. For example, how often they click on recommendations presented by the candidate system as opposed to random recommendations, or those generated by another system. Where it is practical, online evaluation is ideal, as significant differences between online evaluation and offline evaluation have been recorded [11].

More common, however, especially in academic literature, is offline evaluation. In this case, data not used in training is partitioned to be used for validation. The recommender system is trained to predict either ratings on a scale or a binary result (for example, whether or not a user purchases an item). These predictions on the validation dataset can then be compared with the actual results to compute performance metrics.

We first examine the evaluation of user-item recommender systems, as this will be used as a base from which to define metrics for reciprocal systems.

2.5.1 Recommender System Evaluation

Early recommender systems used normalized error as the primary metric of success. This was usually for systems where star ratings of products was common, so the recommender system's objective was to predict a value on a scale, although it can also be used for binary classification. Commonly, this would be the *Root Mean Squared Error* (RMSE), which for N ratings of $r_i i \in 1..N$, and for p_i as the rating for r_i predicted by the recommender system, the RMSE is defined as:

$$RM\bar{S}E = \sqrt{\frac{\sum_{i=1}^{N}(p_i - r_i)^2}{N}} \tag{2.10}$$

Modern evaluations are more likely to use *Precision* and *Recall* as primary metrics of success of the recommender system [4]. In this context, precision is defined as the proportion of recommendations that were successful compared to recommendations made. Recall is the proportion of potentially successful recommendations retrieved out of all potentially successful recommendations. Precision tends to be a more useful metric: users care more if most recommendations are useful than if they saw all possible recommendations. Precision and Recall are illustrated in the general sense in the context of *True Positives*, *False Positives*, *False Negatives* and *True Negatives* in Figure 2.5.

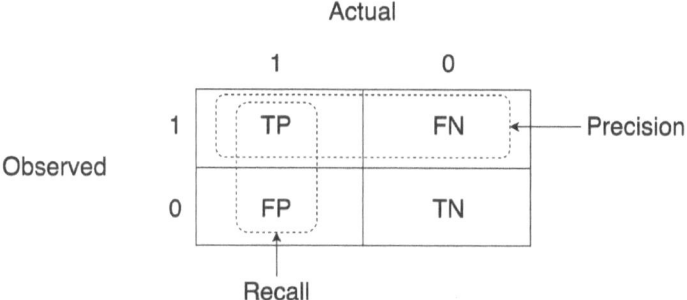

Fig. 2.5: Precision and Recall

Define the set of recommendations made by the system as R and the set of successful recommendations as Rs, then the precision is defined as:

$$Precision = \frac{|Rs|}{|R|} \tag{2.11}$$

We define all possible successful recommendations as the set Ps, then the recall is defined as:

$$Recall = \frac{|Rs|}{|Ps|} \tag{2.12}$$

When evaluating recommender systems, it is most useful to consider these two metrics separately: as stated above, precision tends to be weighted as the more important metric, and separately they give more information about the system. However, it is often possible to tune a system to favour one metric at the expense of the other, and it is not usually desirable to exclude recall entirely: a system that recommends the highest confidence item over and over again is likely to have very high precision, but is not necessarily useful to the user. In order to gain a complete picture of the effectiveness of the system, the metrics are often combined in the literature into the *F1 Score*, which is the harmonic mean of the two scores, and is defined as:

$$F1 = \frac{2 * Precision * Recall}{Precision + Recall} \tag{2.13}$$

A number of other metrics have been proposed for measuring specific elements of recommender systems, such as *Serendipity* [2] - the extent to which a recommender system can offer a user recommendations they would not have searched for themselves. While these are interesting and useful, they have not been applied extensively to reciprocal systems and will therefore not be discussed extensively here.

2.5.2 Reciprocal Recommender System Evaluation

In the case of RRSs, it is not enough to recommend users whom the target user likes. Success depends on mutual preference. In fact, unreciprocated recommendations can discourage users from staying on the service, making them worse than no recommendation at all. Our evaluation methods must therefore reflect this difference [10].

In this domain, the precision and recall functions are modified such that reciprocity is included in the definition of a successful recommendation. We define *Reciprocal Precision* and *Reciprocal Recall* as the metrics by which we measure the performance of RRSs. In the following definitions, RL is the set of users who were recommended to each other and expressed mutual preference, and RN is the set of users who were recommended to each other, and one of them subsequently expressed negative preference. Reciprocal precision is then the proportion of recommended items where the two users expressed mutual preference for each other, defined as:

$$Precision = \frac{|RL|}{|RL| + |RN|} \tag{2.14}$$

And reciprocal recall is the proportion of the total set of potentially successful reciprocal recommendations retrieved by the system. Where P is the set of all observed Matches between users, this is defined as:

$$Recall = \frac{|RL|}{|P|} \tag{2.15}$$

The *Reciprocal F1 Score* can then be defined in the same way as in Section 2.5.1, by taking the harmonic mean of reciprocal precision and recall.

While it is essential to have an understanding of how to evaluate an RRS, this book does not dwell on evaluation data for specific systems beyond discussions of comparative effectiveness. This is because RRSs are in a unusual position in the literature, where none of them have public datasets on which their evaluations can be verified. While it is easy to compare and contrast the (offline) effectiveness of user-item recommender systems based on large public datasets such as the *MovieLens* and *Netflix* datasets, legitimate user privacy concerns prevent RRS environments such as online dating services from releasing datasets. Possibly because of the variance in how information is presented to users, and the diverse demographics of users on these services, implementation of the same algorithm in different contexts has often given very different results. Consider how some services encourage users to make decisions based on a single photo, whereas others have users view a profile with a variety of information before making a decision. Rather than fixating on a few numbers as performance metrics for a given system, we discuss advantages and disadvantages for the algorithms presented. Developers are encouraged to evaluate what kind of RRS would be suitable for their own system based on those advantages and disadvantages, and potentially implement more than one and explore the above performance metrics on their own systems.

References

1. Aggarwal, C.C., et al.: Recommender systems, vol. 1. Springer (2016)
2. Ge, M., Delgado-Battenfeld, C., Jannach, D.: Beyond accuracy: evaluating recommender systems by coverage and serendipity. In: Proceedings of the fourth ACM conference on Recommender systems, pp. 257–260 (2010)
3. Gu, J., Wang, Z., Kuen, J., Ma, L., Shahroudy, A., Shuai, B., Liu, T., Wang, X., Wang, G., Cai, J., et al.: Recent advances in convolutional neural networks. Pattern recognition **77**, 354–377 (2018)
4. Gunawardana, A., Shani, G.: A survey of accuracy evaluation metrics of recommendation tasks. Journal of Machine Learning Research **10**(12) (2009)
5. Hearst, M.A., Dumais, S.T., Osuna, E., Platt, J., Scholkopf, B.: Support vector machines. IEEE Intelligent Systems and their applications **13**(4), 18–28 (1998)
6. Hecht-Nielsen, R.: Theory of the backpropagation neural network. In: Neural networks for perception, pp. 65–93. Elsevier (1992)
7. Herlocker, J.L., Konstan, J.A., Terveen, L.G., Riedl, J.T.: Evaluating collaborative filtering recommender systems. ACM Transactions on Information Systems (TOIS) **22**(1), 5–53 (2004)
8. Lin, T., Wang, Y., Liu, X., Qiu, X.: A survey of transformers. AI open **3**, 111–132 (2022)
9. Neve, J., Palomares, I.: Aggregation strategies in user-to-user reciprocal recommender systems. In: 2019 IEEE International Conference on Systems, Man and Cybernetics (SMC), pp. 4031–4036. IEEE (2019)
10. Pizzato, L., Rej, T., Chung, T., Koprinska, I., Kay, J.: Recon: a reciprocal recommender for online dating. In: Proceedings of the fourth ACM conference on Recommender systems, pp. 207–214 (2010)
11. Shani, G., Gunawardana, A.: Evaluating recommendation systems. Recommender systems handbook pp. 257–297 (2011)

12. Van Meteren, R., Van Someren, M.: Using content-based filtering for recommendation. In: Proceedings of the machine learning in the new information age: MLnet/ECML2000 workshop, vol. 30, pp. 47–56. Barcelona (2000)

13. Xia, P., Liu, B., Sun, Y., Chen, C.: Reciprocal recommendation system for online dating. In: Proceedings of the 2015 IEEE/ACM International Conference on Advances in Social Networks Analysis and Mining 2015, pp. 234–241 (2015)

Chapter 3
Collaborative Filtering

3.1 Introduction

Collaborative filtering uses correlations between user preferences to make predictions. As discussed in Chapter 1, the basic concept is that if Alice likes X, Y and Z, and Kate likes X and Y, we might assume that she has similar preferences to Alice and therefore recommend her Z. The challenge then becomes applying this theory to a very large number of users with potentially very sparse preference data, where these kinds of clear and direct correlations are not always immediately apparent.

Collaborative filtering has been the most successful model of recommendation in the literature. Systems using this method first originated in the 1990s. In particular, *GroupLens*, developed in 1994, used *Nearest Neighbour Collaborative Filtering* to make recommendations for internet news to users based on correlations between them. These algorithms, such as the one described in Section 2.2, find similar users by calculating a similarity metric.

Collaborative filtering advanced significantly in the 2000s with the *Netflix Prize* competition [2]. The company *Netflix* released a large dataset of user movie recommendations in 2006 with large prizes awarded to users who were able to design algorithms to beat the offline accuracy of their own recommendation algorithm. By 2009, the accuracy of algorithms on this dataset had improved by over 10%, with Latent Factor-based algorithms playing a significant role in its advancement [1]. These algorithms make recommendations by calculating higher order properties of preferences and attributes.

Since then, modern machine learning methods have increasingly been incorporated into collaborative filtering for latent factor models. In particular, CNNs have been effective at learning latent factors from data, improving results in online environments. Infrastructure has also contributed to this: more and more websites log granular user actions such as clicking and scrolling, which provides more implicit data from which to understand user preferences.

Collaborative filtering has several advantages over content-based filtering. The most important of these is performance metrics: historically, collaborative filtering

© The Author(s), under exclusive license to Springer Nature Switzerland AG 2025
J. Neve, *Reciprocal Recommender Systems*, SpringerBriefs in Computer Science,
https://doi.org/10.1007/978-3-031-85103-2_3

has consistently been better than content-based filtering at making recommendations based on historical preferences, from offline and online performance metrics. This is equally true in reciprocal environments where, outside of a few specific situations, collaborative filtering algorithms tend to outperform content-based filtering.

A further distinction worth being aware of within collaborative filtering is that between *Memory-Based Collaborative Filtering* and *Model-Based Collaborative Filtering* [5]. Memory-based methods generate recommendations directly from raw preference data. Model-based methods process preference data into a meaningful model, from which recommendations can quickly be generated. As a general rule, model-based methods tend to have better performance, especially on larger datasets, because they are often able to capture higher-order patterns within the data, whereas memory-based methods tend to base their recommendations on pairwise comparisons, and often restrict these to a subset avoid time complexity becoming a problem.

This chapter outlines two types of collaborative filtering which have successfully been applied to reciprocal recommendation with positive results: *Nearest Neighbor Collaborative Filtering* with *RCF* as a case study of memory-based collaborative filtering, and *Latent Factor Models* with *LFRR* as a case study of model-based collaborative filtering.

3.2 Preference Indicators

3.2.1 Preference Extraction

Modern online dating services often require users to exchange binary indicators of preference before they can communicate with each other. These indicators of preference can be implicit or explicit, and are the primary form of data used in the design of collaborative filtering algorithms.

The most important of these indicators of preference are described below. These, along with their formalisations, are used throughout the rest of this book, but particularly critical in collaborative filtering. They are also depicted visually in Figure 3.1.

- A **Like** is a positive, binary, unidirectional indicator of preference. On many online dating services, this is explicitly given as an option to the user when they view another user's profile. Likes are usually a limited resource, awarded after a certain time period or in exchange for money. Users therefore have an incentive to be selective, and can't just Like every user indiscriminately in the hope of getting some responses. Services which do not have explicit Likes can often use a similar binary indicator of preference, for example, sending an initial message. We formalise a Like from user A to user B as $\mathcal{L}(A, B)$.

- A **Match** is a positive, binary, bidirectional indicator of preference. A Match results from the mutual exchange of Likes between two users. On some services, this is presented as a coincidence, and users are not aware of which other users

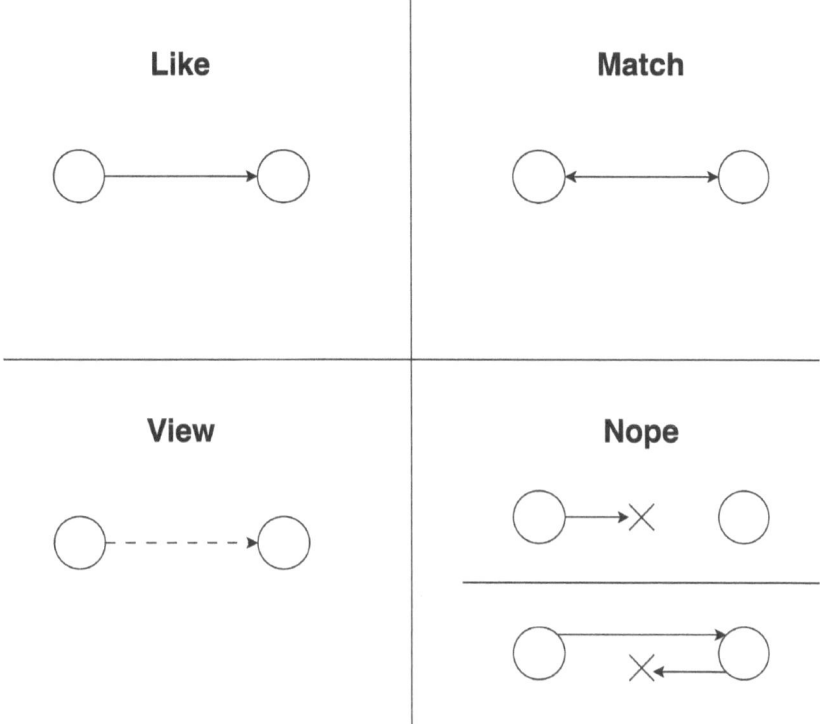

Fig. 3.1: Indicators of Preference in RRSs

have Liked them. Most services, however, show users other users who have Liked them, and give them the option to respond. A Match is a prerequisite for communication such as messaging on many services. We formalise a Match between users A and B as $\mathcal{M}(A, B)$.

- A **Nope** is a negative, binary, unidirectional indicator of preference. How Nopes are presented once again differs depending on the service. Some services give users the explicit option to express a negative or positive preference via, for example, swiping left or right respectively. In services which rely on search, we may need to infer negative preferences implicitly. For example, if Alice is Liked by Bob and, after seeing this, chooses not to respond with a mutual Like, we could consider this an implicit Nope. We formalise a Nope from A to B as $\mathcal{N}(A, B)$.

- A **View** is a binary, unidirectional indicator of preference. A View indicates that one user looked at another user's profile. This is not necessarily a positive or negative indicator, as while a user clicking on another user's profile can indicate some degree of preference, if Alice views Bob's profile but chooses not to send a Like, it can indicate that there was something about Bob's profile that was not appealing to her. Views are, however, important in contexts such as popularity. A low number of Likes with no context could indicate that the user is not appearing in search results often for any reason (for example, because they live in a remote area). A high ratio of Likes to Views usually indicates that the user is appealing in the general sense. We formalise a View from A to B as $\mathcal{V}(A, B)$.

These pieces of information are most commonly used in reciprocal recommendation. In particular, Likes and Nopes are often used in the training process, and Matches are used as a metric of the algorithm's success.

It is worth noting that depending on the service, its operation and what user data the service captures and has permission to use for recommendation, there is various other information that may contribute to user preference. This might be as granular as scrolling patterns through search results, clicks on sub-photos, exchanges of contact details and so on. Because of how service dependent they are, and because they very rarely appear in published literature they will not be referred to explicitly here. However, it is worth being aware that they exist and that it might be valuable to incorporate them into preference score calculations.

Other services besides online dating often have similar forms of binary data where a relationship between two parties is initiated, which we can use in a similar way. For example, on a recruitment service, a CV being sent by a user might be considered a Like, a positive message from the company or an invitation to interview might be considered a Match, whereas a rejection might be considered a Nope.

Modern services collect a wealth of data besides these binary indicators. Before Likes are exchanged, services might log user behaviour such as logins, search criteria, profile views and so on. After Likes have been exchanged, users often exchange messages on the service. It seems like this data should also be incorporated into reciprocal recommender systems, but there are as of yet no examples of it. This is largely because, while this data does provide information as to user intentions, it is more difficult than binary preference indicators to interpret and evaluate. If a Alice views Bob's profile and then doesn't send a Like, we could interpret this as positive preference (Alice liked Bob's photo enough to click) or negative preference (she decided not to send a Like). Similarly, message content after Matching is often conversational, and difficult to interpret positive or negative signals from unless one user says something particularly flattering or abrasive. RRSs tend to stick to the binary signals described above, both because they are clear positive and negative signals, and because they provide points of comparison to other, existing algorithms.

3.2.2 Preference Representation

Having established our positive and negative preference indicators, we can create a *Preference Matrix*. Early reciprocal recommender systems are based on services which split the user base into two groups, often *Male* and *Female*, and create two unidirectional preference matrices. However, as services increasingly aim to cater to all sexual orientations, this is becoming less useful and less common. This chapter will therefore assume a single preference matrix, where any user can express preference

Likes	$P_{x\to\text{Alice}}$	$P_{x\to\text{Bob}}$	$P_{x\to\text{Charlie}}$	$P_{x\to\text{Diana}}$
$P_{\text{Alice}\to y}$	X	1	0	0
$P_{\text{Bob}\to y}$	1	X	1	1
$P_{\text{Charlie}\to y}$	0	0	X	1
$P_{\text{Diana}\to y}$	0	1	1	X

Table 3.1: An example preference matrix

for any other user. This is easy to adapt to services where there is a clear division between users, whether this be dating services catering to specific orientations, or other applications of RRSs such as recruitment which divides users into companies and candidates. For a user Alice, we can exclude users with whom she cannot match with for any reason from the recommendation generation process.

As in the previous chapter, we use $P_{\text{Alice}\to\text{Bob}}$ to indicate the number which represents the preference of Alice for Bob. Table 3.1 shows an example preference matrix, where 1 represents a Like, 0 represents a Nope and a blank cell represents that the user has not expressed a preference. The table is completed for the purposes of explaining the algorithms in this chapter, but on most services these matrices are both sparse and extremely large.

While these binary indicators are commonly used as part of algorithms, it is worth keeping in mind that they usually represent the best information offered to us by the service, but do not necessarily represent the objective of users, which will usually vary depending on the user and the service. For example, on an online dating service, a user's objective might be to find a serious long-term partner. Unfortunately, whether or not a Match led to this is often not information that the service has, so we usually have to make do with what we do have. We hope that our proxy for success (i.e. increasing the number of Matches) also increases the number of users who are able to achieve their ultimate objectives. It is worth bearing in mind that becoming laser focused on small percentage increases in the number of Likes and Matches, while potentially satisfying from an algorithmic point of view, might not improve the user experience.

3.3 Nearest Neighbor Collaborative Filtering

Neighborhood methods of collaborative filtering determine similarity between users by using a correlation coefficient between users' preference histories. Having determined k similar users to the target user, we can make recommendations based on what those similar users liked. Section 2.2 introduced a neighborhood model commonly used in user-item recommendation environments and the *Pearson Correlation Coefficient*. Here, we introduce a commonly used adaptation for reciprocal environments, along with some extensions to it.

3.3.1 RCF: A Nearest Neighbor Recommender System

RCF (Reciprocal Collaborative Filtering) (Xia et al. [6]) uses a nearest neighbor method to perform collaborative filtering. It was the first collaborative filtering algorithm to significantly outperform content-based systems, and has been successfully implemented a number of times since and used as a baseline for other algorithms.

RCF calculates the unidirectional preference for two users individually. It then combines these two unidirectional preference scores into a single reciprocal preference score which can be used for recommendation. It bases the unidirectional preference calculation on similarity scores defined between two users. The algorithm takes the following steps to calculate the preference of Alice for Bob:

1. Find the set of users who have Liked Bob.
2. Calculate the similarities between Alice and each of the users who have Liked Bob. We do this as follows. For each user who has liked Bob, such as Clara:
   (a) Find the sets of users who Alice and Clara have Liked.
   (b) Calculate a similarity metric between the two sets of users.
3. Calculate the average similarity between Alice and the users who have Liked Bob. This is considered Alice's preference for Bob.

The similarity calculation described in step 2(b) is a particularly important part of the algorithm. RCF defines user similarity using the *Jaccard Similarity Coefficient*, which we take a short detour to define. This is similar to the Pearson Correlation Coefficient introduced in Section 2.2, but more appropriate for this context. The Pearson Correlation Coefficient used for calculating linear correlations, where user preferences are expressed on a scale such as a 1-5 star rating. The Jaccard Similarity Coefficient, which measures the similarity between two sets, fits better with the binary data such as the presence or absence of a Like or a Message. This tends to be useful in reciprocal environments, where users are rarely encouraged to rate each other on a linear scale, and indicators of preference such a Likes, clicks and messages are usually binary.

For two sets X and Y, the Jaccard Similarity Coefficient $J(X, Y)$ is defined as:

$$J(X, Y) = \frac{|X \cap Y|}{|X \cup Y|} \tag{3.1}$$

Intuitively, this is the number of items the sets have in common, divided by the total number of items. For example, if $X = \{1, 3, 4, 6, 8\}$ and $Y = \{1, 3, 4, 5, 7\}$ then:

$$J(X, Y) = \frac{|\{1, 3, 4\}|}{|\{1, 3, 4, 5, 6, 7, 8\}|} = \frac{3}{7} \tag{3.2}$$

RCF adapts the Jaccard Similarity Coefficient to create a similarity metric between the preferences of two users in an RRS environment as follows. In order to compute similarity, we first define two key sets. First, let $\mathcal{L}(X, Y)$ mean X has sent a Like to Y. Then, we define the set of users who have received a Like from X:

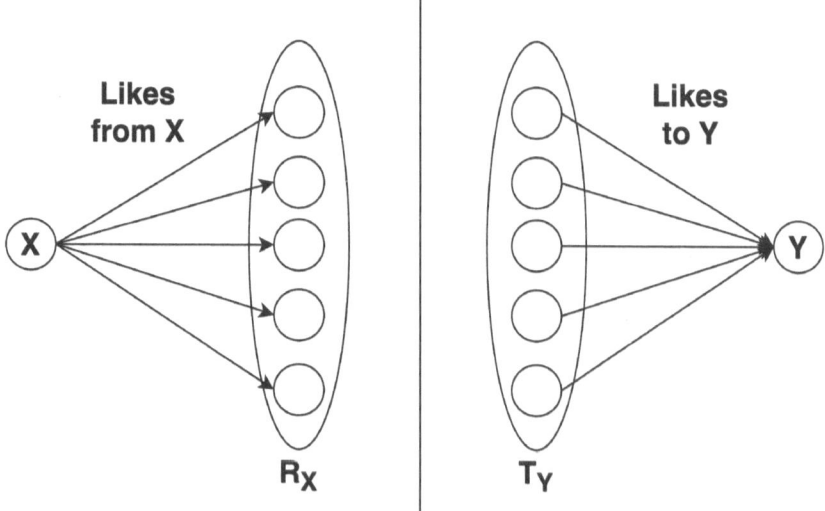

Fig. 3.2: R_X and T_Y sets used for RCF

$$R_X = \{Y \in U : \mathcal{L}(X,Y)\} \tag{3.3}$$

Second, we define the set of users who have Liked Y as:

$$T_Y = \{X \in U : \mathcal{L}(X,Y)\} \tag{3.4}$$

The difference between these two sets is not always very intuitive from the formal definition, but they are displayed in Figure 3.2 for reference. The left side of the image shows R_X, and the right side T_Y. Then, we define the similarity between two users X and Y as:

$$S_{X,Y} = \frac{|R_X \cap R_Y|}{|R_X \cup R_Y|} \tag{3.5}$$

The similarity score increases with the proportion of Liked users that X and Y have in common. At the edge case where both R_X and R_Y are zero, $S_{X,Y}$ is undefined. Neither user has Liked anyone, so we have no information about their preferences to calculate a similarity. It is worth being aware of this edge case, because if these pairs are not skipped during computation, a naive implementation might result in division by zero.

We can use similarity scores with other scores to make predictions about the unidirectional preferences from Alice to Bob. First, we find the set T_B of users who have Liked B. Then we compute the similarity between A and each of those other users using the Jaccard Similarity Coefficient as defined above. Finally, to calculate Alice's preference for Bob $P_{A \to B}$, we sum those similarity scores, and normalize by the number of users who have Liked B:

$$P_{A \to B} = \frac{1}{|T_B|} \sum_{x \in T_B} S_{A,x} \qquad (3.6)$$

This computation represents the average similarity between A and the other users who have Liked B.

We can calculate Bob's preference for Alice $P_{B \to A}$ using the same method. Finally, the Harmonic Mean is used to calculate the reciprocal preference score $P_{A \leftrightarrow B}$ which represents the reciprocal preference between Alice and Bob:

$$P_{A \leftrightarrow B} = \frac{2 P_{A \to B} P_{B \to A}}{P_{A \to B} + P_{B \to A}} \qquad (3.7)$$

As discussed in Section 2.4 on preference aggregation, the Harmonic Mean is often used to penalise a large difference between the unidirectional preference scores, as one low score often means that the chance of reciprocal preference is low, even if the other score being aggregated is very high.

Likes	$P_{x \to \mathbf{Alice}}$	$P_{x \to \mathbf{Bob}}$	$P_{x \to \mathbf{Charlie}}$	$P_{x \to \mathbf{Diana}}$	$P_{x \to \mathbf{Edward}}$
$P_{\mathbf{Alice} \to y}$	X	?	1	1	1
$P_{\mathbf{Bob} \to y}$	?	X	0	1	0
$P_{\mathbf{Charlie} \to y}$	1	0	X	1	1
$P_{\mathbf{Diana} \to y}$	0	1	0	X	1
$P_{\mathbf{Edward} \to y}$	1	1	1	0	X

Table 3.2: A preference matrix for RCF

As an example, we can attempt to predict the preferences of Alice and Bob for each other in Table 3.2 using RCF. Firstly, to predict $P_{\mathrm{Alice} \to \mathrm{Bob}}$, we compare the users who have Liked Bob to Alice. Those users are Diana and Edward. We then calculate the similarity between Alice and Diana, and Alice and Edward.

Alice's Liked set is {Charlie, Diana, Edward} while Diana's Liked set is {Bob, Edward}. Their Similarity is the size of intersection of those two sets over their union:

$$S_{\mathrm{Alice, Diana}} = \frac{|\{\mathrm{Edward}\}|}{|\{\mathrm{Bob, Charlie, Diana, Edward}\}|} = \frac{1}{4} \qquad (3.8)$$

For the similarity between Alice and Edward, Edward's Liked set is {Alice, Bob, Charlie} and their Similarity is then:

$$S_{\mathrm{Alice, Edward}} = \frac{|\{\mathrm{Charlie}\}|}{|\{\mathrm{Alice, Bob, Charlie, Diana, Edward}\}|} = \frac{1}{5} \qquad (3.9)$$

The preference of Alice for Bob can then be calculated as the average of the similarities:

$$P_{\text{Alice}\rightarrow\text{Bob}} = \frac{0.25 + 0.2}{2} = 0.225 \tag{3.10}$$

To calculate the reverse preference $P_{\text{Bob}\rightarrow\text{Alice}}$, we perform similar calculations to calculate the similarities between Bob and the set of users who have Liked Alice, Charlie and Edward. The similarities are $\frac{1}{3}$ and 0 respectively, which we then average in the same way to give:

$$P_{\text{Bob}\rightarrow\text{Alice}} = \frac{0.333 + 0}{2} = 0.167 \tag{3.11}$$

Finally, we use the harmonic mean to calculate the reciprocal preference:

$$P_{\text{Alice}\leftrightarrow\text{Bob}} = \frac{2 \cdot 0.225 \cdot 0.167}{0.225 + 0.167} = 0.192 \tag{3.12}$$

Low scores are common for RCF, especially in larger services where users will often encounter different subsets of other users, and the chance of one user having a very high proportion of users in common with another is low. The importance of the reciprocal score lies in the comparison with other scores, which allows us to rank recommendation candidates relative to each other, and should not be viewed as an even distribution between 0 and 1.

3.3.2 RCF Implementation

This section describes a sample implementation of the RCF algorithm in Python. The code follows the logic described above exactly. For readers whose have backgrounds in coding, this may be slightly easier to follow.

We first define a function to calculate the Jaccard Similarity between two users. For computing the similarity between A and B, set1 is the set of users Liked by A, and set2 is the set liked by B. We also handle the divide-by-zero edge case.

```
def jaccard_similarity(set1, set2):
    if not set1 and not set2:
        return 0  # undefined similarity when both sets are empty
    return len(set1 & set2) / len(set1 | set2)
```

We then define a function to calculate the preference score $P_{A\rightarrow B}$. For each user who has Liked B, we calculate the Jaccard Similarity between them and A based on the comparison between their Liked users and A's Liked users. We add them to a list of similarities, which is summed at the end and used to calculate the overall similarity score. If there are no users who have liked B, we return zero to avoid division by zero errors.

```
1  def preference_score(liked_B, user_likes, A):
2      similarities = []
3
4      for user in liked_B:
5          similarity = jaccard_similarity(user_likes[A], user_likes
               [user])
6          similarities.append(similarity)
7
8      if len(similarities) == 0:
9          return 0
10
11     return sum(similarities) / len(similarities)
```

Next, we define the harmonic mean function which handles the preference aggregation, as a utility for the following function. Once again, we handle the case when the two input values are zero to avoid a divide by zero.

```
1  def harmonic_mean(x, y):
2      if x + y == 0:
3          return 0
4      return (2 * x * y) / (x + y)
```

Finally, we use these functions to calculate the reciprocal preference, by first getting the users who have Liked each user as input for the following function, and then using them to calculate the two unidirectional preference scores $P_{A \to B}$ and $P_{B \to A}$. We aggregate these two scores using the harmonic mean, and return the calculated reciprocal preference score. liked[A] is the set of users who have sent Likes to user A.

```
1  def reciprocal_preference(A, B, user_likes, liked):
2      # Get the set of users who liked A and B
3      liked_A = liked[A]
4      liked_B = liked[B]
5
6      # Calculate unidirectional preferences
7      P_A_to_B = preference_score(liked_B, user_likes, A)
8      P_B_to_A = preference_score(liked_A, user_likes, B)
9
10     # Use harmonic mean to calculate the reciprocal preference
11     return harmonic_mean(P_A_to_B, P_B_to_A)
```

Having defined all the necessary functions, we input our data. Here, we use the data from Table 3.2 to define user_likes. We then define liked as a defaultdict from the collections package. This allows us to efficiently access the users who have sent a Like to each user (i.e. user_likes gives us easy access to R_X, while liked gives us easy access to T_Y).

```
 1  user_likes = {
 2      'Alice': {'Charlie', 'Diana', 'Edward'},
 3      'Bob': {'Diana'},
 4      'Charlie': {'Alice', 'Diana', 'Edward'},
 5      'Diana': {'Bob', 'Edward'},
 6      'Edward': {'Alice', 'Bob', 'Charlie'}
 7  }
 8
 9  liked = defaultdict(set)
10  for user, likes in user_likes.items():
11      for liked_user in likes:
12          liked[liked_user].add(user)
```

Finally, we can calculate the reciprocal preference between two named users by calling the function with our sample data.

```
 1  reciprocal_preference('Alice', 'Bob', user_likes, liked)
```

Interested readers might consider inputting this algorithm and confirming that the result is the same as in Equation 3.12 (0.191).

3.3.3 Time Complexity of RCF

Where collaborative filtering algorithms are concerned, complexity is a particularly important issue. This is because in generating recommendations for users, a naively implemented collaborative filtering algorithm often scales very poorly with the amount of data on the service. As background, this section briefly discusses the time complexity of kNN algorithms for conventional user-item recommender systems. We then move on to examining the time complexity of RCF specifically.

Imagine a user-item recommender system with n users and m items, where we wish to implement a kNN recommender system based on the Pearson Correlation Coefficient as described in Section 2.2. If we want to compute recommendations for every user on the service, we need to perform three sequential steps:

1. **The similarity computation** involves comparing every user with every other user, which is n^2 comparisons. Computing the PCC over the items rated by each pair of users is a further m operations, so the overall time complexity of this step is $O(n^2 \cdot m)$.
2. **Sorting the similarities** to extract k highest similar users using a standard sorting algorithm is $O(n \log(n))$ operations. We must do this for every user, so this step is $O(n^2 \log(n))$ operations.
3. **To predict recommendations**, we aggregate the ratings for m items for k users. Predicting recommendations for every user on the service is therefore complexity $O(n \cdot k \cdot m)$.

The final complexity of generating recommendations is therefore:

$$O(n^2 \cdot m + n^2 \log(n) + n \cdot k \cdot m) \tag{3.13}$$

The $O(n^2 \cdot m)$ term will usually dominate this calculation. For a smaller shopping service with thousands of users and hundreds of items, this is feasible to compute. However, as user numbers increase, the naive implementation becomes increasingly infeasible.

As an example, we take a service with 1000 items and one million users. It is a reasonable assumption that the PCC calculation, which is based on the dot product, would take around 100 microseconds (i.e. 0.0001 seconds) for one pair of users on modern hardware. In a service with n users there are $\frac{n(n-1)}{2}$ possible pairs of users, which we approximate to $\frac{n^2}{2}$ for simplicity of presentation. Then, to calculate similarity between all pairs of users, the time for all pairs of users is:

$$\frac{1000000^2}{2} \times 0.0001 = 5 \times 10^{11} \times 0.0001 \approx 579 \text{days} \tag{3.14}$$

Large shopping services such as Amazon have many more than a million users and a thousand products, so even with better technology, this is a problem that we need more efficient algorithms to solve.

The time complexity calculation for RCF is similar to this, but has a little more nuance to it. In RCF, to compute the preference of Alice for Bob $P_{A \to B}$, we calculate the Jacard Similarity between Alice and each of the users in the set T_B, the users who have Liked Bob. In the worst case for RCF, every user on the service has Liked every other user. We would therefore need to compare Alice to n different users to compute a single preference score $P_{A \to B}$ for Bob. A straightforward implementation of RCF would run this calculation for all of the approximately $\frac{n^2}{2}$ pairs of users, so the time complexity is $O(n^3)$.

Practically speaking, however, the preference matrix tends to be very sparse, and the number of Likes sent by one user is generally a small fraction of the total users. For a more useful upper bound on the number of operations required, where T_Y is the set of users who have Liked Y, we define $\max(|T_Y|)$ as the largest number of times any user on the service has been Liked. We need to run this calculation for all $\frac{n^2}{2}$ pairs of users which gives us a complexity of:

$$O(n^2 \cdot \max(|T_Y|)) \tag{3.15}$$

Depending on the size of the dating service, $\max(|T_Y|)$ might be anywhere from dozens to thousands of times, but is rarely higher than this. This gives us a similar complexity to our original kNN user-item algorithm. We do fewer similarity calculations per user, but because we are recommending users instead of items, we need to run the preference calculation between all pairs of users, and are therefore not bound by the smaller number of items m. It becomes similarly difficult to use once users pass the hundreds of thousands, which makes it infeasible without modification for larger RRS environments. The most popular online dating services have more than 50 million registered users.

3.3.4 Strengths and Weaknesses of Neighbourhood Models

Neighbourhood models were the earliest form of collaborative filtering models, and have shown positive results in a number of environments. They often outperform content-based methods in user-item environments, and this has also generally been the case in RRS environments, with RCF and modified versions of it consistently producing better experimental results than content-based recommender systems.

A significant advantage of Neighbourhood Models is their ease of implementation. As we saw in Section 3.3.2, we can implement a working version of RCF in around 30 lines of code. For smaller services, with users in the thousands, the precision may not be significantly different from more complex models, and they are therefore a very good choice for a new or growing service requiring reciprocal recommendations to implement.

Model-based collaborative filtering algorithms require training to generate recommendations. While these models may allow us to make more accurate recommendations, especially in offline testing on a static dataset, there are significant costs associated with this. Models often require periodic retraining or suffer drops in accuracy as they get out of date. They may also not be as effective at making recommendations for users new to the service if they were not part of the training dataset. If the service has a sufficiently small number of users, neighbourhood models can allow us to generate recommendations in real time.

As discussed in Section 3.3.3, the main weakness of neighbourhood models is their time complexity. As services grow past hundreds of thousands to millions of users, naive implementations will quickly pass the point where generating recommendations for large numbers of users is feasible. While there are methods of reducing the time complexity such as by clustering users and running the algorithm on those clusters, these methods have not demonstrated the same level of accuracy. In particular, in sparse datasets (which RRS environments often are), reducing complexity by clustering users can make it difficult to find suitable neighbours. They also increase the complexity of the implementation, often to the point where they lose their simplicity advantage over model-based methods.

Neighbourhood models have also been shown to be limited in their ability to capture complex patterns. kNN algorithms base their recommendations on the assumption that preference can be predicted from a straightforward similarity computation between a limited number of k users, which potentially excludes all of the data outside of those k users. Where preference is complex (as it certainly is in reciprocal environments), model-based approaches, which make preferences based on calculations across the entire dataset, are often more accurate.

3.4 Latent Factor Models

The previous section focused on neighbourhood-based models of collaborative filtering in reciprocal environments. We looked at some of the advantages of neighbourhood models, and also some of the disadvantages. This chapter focuses on collaborative filtering based on latent factor models, which improve on some of those disadvantages.

If a friend asks us to recommend a movie, we don't usually ask them to list every movie they have ever seen and their rating for that movie. This would be time-consuming for them, and our assessment of their actual preferences might get lost in the details of individual movies. Instead, we ask them more general questions such as what genres the like, what actors they like and so on. Similarly, if a friend asks us to set them up on a blind date, we might ask what kind of appearance they like, what personality traits are they attracted to and so on. This general information is usually more helpful in making recommendations than a large amount very specific information.

Latent factor models aim to identify general trends by extracting a small set of key attributes from the data. Using historical preferences, we compute which attributes users possess and prefer. Because the reduction is based on a trained model and not intuitive, we do not expect these latent factors to correspond exactly to human concepts as concrete as genre, but the concept is the same: to reduce a large, sparse user preference matrix to a few descriptive numbers.

3.4.1 Matrix Factorization

The matrix factorisation process is central to how latent factor models are constructed. Factorising a matrix means dividing it into two or more lower dimensional matrices which, when multiplied together, give the original matrix. In recommender systems, this dimensionality reduction process is usually approximated with machine learning techniques. Before that is explained in detail, however, it is useful to briefly explore the linear algebra that our recommender system methods originated from.

The technique traditionally used to perform matrix factorization is the *Singular Value Decomposition* (SVD). Let our original preference matrix be $\mathcal{R}$. The SVD breaks down the matrix $\mathcal{R}$ into three matrices which perfectly reconstruct it:

$$\mathcal{R} = U\Sigma V^T \tag{3.16}$$

The pen-and-paper method of factorising $\mathcal{R}$ into these three matrices involves calculating *eigenvalues* and *eigenvectors* from the *gram matrix* of $\mathcal{R}$ defined as $(\mathcal{R}\mathcal{R}^T)$, which give us Σ and V respectively, and then using these and the original matrix $\mathcal{R}$ to calculate U. The matrices U and V contain latent factors, and the matrix Σ is a diagonal matrix which contains information about the strength of those latent factors. A thorough treatment of this method from the linear algebra perspective is outside the scope of this book, but interested readers are encouraged to look it up, as it provides useful (although not essential) background for the machine learning factorisation process.

Using the SVD directly gives us complete information about the matrix, but the technique is rarely used directly as part of recommendation. This is because the SVD requires a complete matrix. The preference matrix is usually extremely sparse, and has many unknown values. However, the SVD can only be performed on a matrix which is completely filled. Filling the missing values with a neutral value such as 0.5 does not generally work very well, as the matrix is so sparse that the neutral values dominate the computation, and the resulting U and V do not become very accurate predictors of preferences.

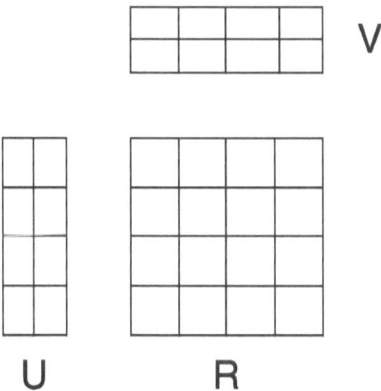

Fig. 3.3: Matrices U, V and $\mathcal{R}$

Instead, we use machine learning methods to train a U and V from our sparse preference matrix $\mathcal{R}$, which can then be used to predict any unknown value. Then, for the purposes of collaborative filtering, our objective is to decompose this into two matrices U and V which *approximate* the full version of $\mathcal{R}$, including missing values. We call this approximation $\hat{\mathcal{R}}$, such that:

$$\hat{\mathcal{R}} = UV^T \tag{3.17}$$

In this equation, U ($n \times k$ matrix) represents user preferences and V ($k \times n$ matrix) represents user properties. This is visualised in Figure 3.3.

This matrix multiplication implies that the cell which represents $P_{A \to B}$ in the preference matrix is computed by the dot product of two vectors, the row A of U (U_A) which represents the preferences of A, and the column B of V (V_B^T), which represents the properties of B:

$$P_{A \to B} = U_A \cdot V_B^T \tag{3.18}$$

This is visualised in Figure 3.4. This is particularly convenient for recommendation, where our primary objective is to predict unknown preference values. This also gives us an immediate advantage over neighbourhood models in terms of time complexity, where calculating a single preference no longer requires similarity calculations and comparisons between users; just a single vector dot product calculation.

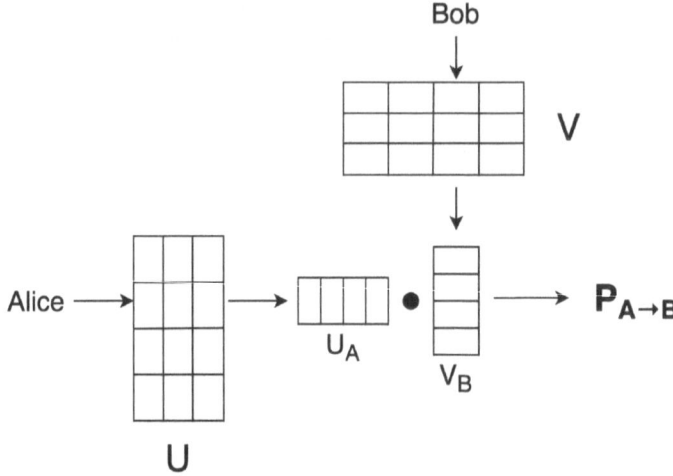

Fig. 3.4: Calculating $P_{A \to B}$ from U_A and V_B

3.4.2 LFRR: Latent Factors for RRSs

With an understanding of what matrix factorisation is, we now move on to exactly how we extract U an V from $\mathcal{R}$ in the context of a reciprocal recommender system. For the purposes of this, we examine *LFRR* (Latent Factor Reciprocal Recommender) (Neve et al. [4]), which was the first RRS to adapt latent factor methods to reciprocal environments.

As the previous section described, starting from the original sparse preference matrix $\mathcal{R}$ with dimensions $n \times n$ (where n is the number of users), we want to find two matrices U and V of dimensions $n \times k$ where k is the number of latent factors. These matrices are such that their product UV^T gives us a new matrix $\hat{\mathcal{R}}$ where $\hat{\mathcal{R}}_{AB}$ contains the predicted preference of Alice for Bob $P_{A \to B}$. We do not have to perform the entire matrix multiplication for this result; we can find it from the dot product of the vectors from the corresponding rows of U and V:

$$\hat{\mathcal{R}}_{AB} = U_A \cdot V_B^T = \sum_{x=1}^{k} U_{Ax} \cdot V_{Bx} \qquad (3.19)$$

This formula is particularly important to the training process, because it allows us to compute an *error* for individual cells. If we want to improve our latent factor matrices U and V, we first calculate the error as the difference between a previously observed interaction $\mathcal{R}_{AB}$ (which might be, for example, a 1 or a 0 depending on whether a user sent a Like or a Nope), and the approximated value from our latent factor matrices $\hat{\mathcal{R}}_{AB}$:

$$e_{AB} = |\mathcal{R}_{AB} - \hat{\mathcal{R}}_{AB}| \qquad (3.20)$$

From our error function, we then define a *loss function* as the sum of the squared errors across all pairs of users (A, B) in $\mathcal{R}$. Let $\mathcal{R}_O$ denote the set of observed entries in R, then:

$$L = \sum_{(A,B) \in \mathcal{R}_O} (R_{AB} - U_A \cdot V_B^T)^2 + \lambda(\|U\|^2 + \|V\|^2) \qquad (3.21)$$

We add the λ term (the sum of the squared magnitudes of U and V) as a *regularisation parameter* to reduce overfitting. It penalises (i.e. increases the loss for) very large latent factors, which are usually the result of overfitting to noise.

Having defined the loss function, LFRR updates latent factors using *Stochastic Gradient Descent* (SGD). SGD considers the loss function as a surface, and aims to reduce the error by computing the gradient of the loss function, and then taking small steps down the gradient in the negative direction, with the objective of finding the minimum error over observed ratings. The result of this will be a U and V which can accurately predict missing values.

Formally, to train a model using SGD with a parameter θ, and we want to update that parameter, the generalised update rule is:

$$\theta \leftarrow \theta - \eta \cdot \nabla_\theta L \qquad (3.22)$$

Where η is the learning rate (i.e. the size of the step we need to take) and $\nabla_\theta L$ is the gradient of the loss function with respect to the parameter θ

In order to determine the gradient of the loss function so we can travel down it in a negative direction, we compute its derivative. In this case, the loss function is based on two variables which we want to update, U and V, so we compute its partial derivatives with respect to those variables. As these are symmetrical, we use U as an example, from which the update rule of V will be intuitive.

The differential of Equation 3.21 with respect to U is as follows:

$$\begin{aligned} &\frac{\partial}{\partial U_A}((R_{AB} - U_A \cdot V_B^T)^2 + \lambda(\|U\|^2 + \|V\|^2)) \\ &= -2V_B \cdot (R_{AB} - U_A \cdot V_B^T) + 2\lambda\|U\| \\ &= -2[V_B \cdot (R_{AB} - U_A \cdot V_B^T) - \lambda\|U\|] \end{aligned} \qquad (3.23)$$

We can then plug this into the generalised update rule described in Equation 3.22 to get the update rule for U_A. The -2 term is factored into the learning rate η:

$$U_A \leftarrow U_A + \eta \cdot [V_B \cdot (R_{AB} - U_A \cdot V_B^T) - \lambda\|U\|] \qquad (3.24)$$

The equivalent process of finding the partial differential with respect to V gives a similar update rule for V_B:

$$V_B \leftarrow V_B + \eta \cdot [U_A \cdot (R_{AB} - U_A \cdot V_B^T) - \lambda\|V\|] \qquad (3.25)$$

The algorithm itself begins with randomised matrices U and V, with values selected in the range $[0, 1]$. We then iterate repeatedly through known values in $\mathcal{R}_O$, making updates using the update rules for U and V based on the error. Each iteration is known as an *epoch*, and is expected to reduce the loss L. When the loss stops improving, we have reached a minimum value of the loss function, and hopefully a U and V capable of predicting values in $\hat{\mathcal{R}}$.

LFRR then uses U and V for prediction in a similar way to previous algorithms. Using the vectors for user preference for Alice and Bob U_A and U_B and the values for user properties V_B^T and V_A^T, we can predict $P_{A \to B}$ and $P_{B \to A}$ using Equation 3.18. We use the Harmonic Mean to combine these two unidirectional preferences into a single reciprocal preference:

$$P_{A \leftrightarrow B} = \frac{2P_{A \to B}P_{B \to A}}{P_{A \to B} + P_{B \to A}} \tag{3.26}$$

The hyperparameters η, k and λ are determined experimentally from results on a test dataset. However, we discuss them briefly here. The learning rate η is generally initially set as a small number (0.01 or 0.001 is a common choice) and then adjusted based on how the loss function is changing with training. If it is too small, the model takes a very long time to improve; if it is too large, the model struggles to find the minimum due to overshooting it. The regularization parameter, λ, depends on the situation, but a value such as 0.1 is not an uncommon starting point. If it is too large, the model fails to fit to details in the data; if it is too small, the model starts fitting too much to noise. Finally, k, the number of latent factors has a similar impact to λ. A small k is unable to capture complex patterns, a larger k takes longer to train, and risks overfitting to the dataset (a sufficiently large k will be able to reproduce the training matrix exactly). A value between 10 and 50 is usually a good starting point.

3.4.3 Time Complexity of LFRR

The time complexity of LFRR is a simpler calculation than that of RCF. RCF uses a number of steps to compute similarity in memory based on who users have Liked and been Liked by. LFRR, on the other hand, has two steps, both with simple complexity. For a service with n users:

1. **Training** is done by repeatedly running through the known preference expressions in an $n \times n$ matrix. The time complexity of this scales with the size of the matrix, $O(n^2)$ occurring when we know the values of every item in the matrix. However, as the matrix is usually very sparse, usually with significantly fewer than 1% of the cells filled, it is often much less than this.
2. **Prediction** consists of a vector dot product for predicting a single preference. As the dot product is $O(1)$, if we wish to predict preferences for all users in the $n \times n$ matrix for every other user, this also scales with $O(n^2)$.

Our final time complexity is $O(n^2)$, which makes it more efficient than the $O(n^2 \cdot \max(|T_Y|))$ which we calculated for RCF [3].

However, this worst-case calculation hides a certain amount of the extra processing required by RCF in practical situations. The preference calculations for RCF are based on relatively small latent factor matrices, which can often be loaded into memory, so calculating a large number of preferences in order to make recommendations is often efficient, and is also easy to parallelise. However, the RCF calculation between Alice and Bob is based on the users who have Liked Bob and the users whom they in turn have Liked. This data often has to be loaded from a database, which becomes a bottleneck for preference prediction. Where RCF and LFRR have been directly compared, the same preference predictions that were computable in seconds with LFRR have taken hours using RCF.

3.5 Choosing a Collaborative Filtering Algorithm

In this chapter, we looked at two algorithms for collaborative filtering. A number of other algorithms have been designed, all with essentially the same objective: given A and B, predict $P_{A \leftrightarrow B}$ from correlations between user preferences. We finish this chapter with a discussion of how to choose which system to use.

There are three main factors to balance:

1. The difficulty of implementation of the algorithm. An algorithm such as RCF is relatively easy to implement, and can be done quickly. A new service with fewer development resources might not get much initial benefit over using more complex models such as latent factor models, so implementing a simple kNN style recommender is likely to be a good choice, compared to larger services with dedicated specialists working on recommendation.
2. The time complexity scaling of the algorithm. Nearest neighbour algorithms tend not to scale well without modification to large datasets. Where we have an online dating service with millions of users, latent factor models are likely to be the better choice, as nearest neighbour algorithms begin to take unreasonable amounts of time to generate recommendations for large numbers of users.
3. The training resources available. While generating recommendations for an entire service might be more time consuming using a memory-based method, model-based methods require training, and an infrastructure capable of retrieving recommendations from a saved model and updating that model. For a service that is accessed less frequently, or only requires recommendations to be generated for a subset of users, a memory-based method may be more desirable.

As many of these factors are service dependent, designers are encouraged to implement and test multiple models using the metrics described in Section 2.5, and evaluate their effectiveness and efficiency before making a final decision.

References

1. Bell, R.M., Koren, Y.: Lessons from the netflix prize challenge. Acm Sigkdd Explorations Newsletter **9**(2), 75–79 (2007)
2. Bennett, J., Lanning, S., et al.: The netflix prize. In: Proceedings of KDD cup and workshop, vol. 2007, p. 35. New York (2007)
3. Kleinerman, A., Rosenfeld, A., Ricci, F., Kraus, S.: Optimally balancing receiver and recommended users' importance in reciprocal recommender systems. In: Proceedings of the 12th ACM Conference on Recommender Systems, pp. 131–139 (2018)
4. Neve, J., Palomares, I.: Latent factor models and aggregation operators for collaborative filtering in reciprocal recommender systems. In: Proceedings of the 13th ACM conference on recommender systems, pp. 219–227 (2019)
5. Su, X., Khoshgoftaar, T.M.: A survey of collaborative filtering techniques. Advances in Artificial Intelligence **2009**, 1–19 (2009)
6. Xia, P., Liu, B., Sun, Y., Chen, C.: Reciprocal recommendation system for online dating. In: Proceedings of the 2015 IEEE/ACM International Conference on Advances in Social Networks Analysis and Mining 2015, pp. 234–241 (2015)

Chapter 4
Content-Based Filtering

4.1 Introduction

In the context of Reciprocal Recommendation, content-based filtering uses a user's historical preference for other users' profiles to calculate their preferences. This is often an effective way of making recommendations, especially in the context of online dating. Users fill in detailed profiles when they sign up, and it is to their advantage to do so: the quality of a user's profile will often determine how much attention they receive from other users.

In the general case, evidence has shown content-based filtering to be less effective than collaborative filtering [3]. This also holds true in the case of online dating where, using the offline testing methods preferred in the literature, collaborative filtering methods tend to outperform content-based filtering methods [10]. However, they are still valuable to study for two reasons. Firstly, content-based filtering systems tend to be simpler to efficiently implement than collaborative filtering systems, and might therefore be preferred, especially on services operated with fewer engineers, for simplicity reasons. Secondly, a sophisticated content-based filtering system can often be very successful in certain situations, and in particular tend to outperform collaborative filtering in cold-start scenarios [13], and therefore are often used as components of a hybrid system.

User profiles in reciprocal environments are generally comprised of three different types of data: categorical data (for example, age, body type, salary), a freetext section where users will describe both themselves and what they might be looking for, and one or more photographs of the user. For ethical and data protection reasons, very little information about which of these factors is the most important has been made public by modern dating services. Some early experiments involving photos were, however, made public, and indicated that the user's photo was the greatest predictor of attraction [1].

Early reciprocal systems were content-based filtering, and focused on the categorical data. Some research has indicated that categorical data is not the main

---

[1] https://www.theguardian.com/technology/2014/jul/29/okcupid-experiment-human-beings-dating

© The Author(s), under exclusive license to Springer Nature Switzerland AG 2025
J. Neve, *Reciprocal Recommender Systems*, SpringerBriefs in Computer Science,
https://doi.org/10.1007/978-3-031-85103-2_4

basis for users' choice of partner, and these early systems were therefore not always highly accurate, but they did provide an important baseline for future systems to build on [9]. They are also easy to implement and to explain, where RRSs based on photos or freetext often involve complex machine learning models, which requires significantly more training overhead and generally aren't explainable.

This chapter outlines several methods of making reciprocal recommendations using content-based filtering in online dating. We start with the earliest example of an online dating RRS, *RECON* [11], which uses historical preference of users to determine user preferences for individual categories in another user's profile. User preferences for individual profile categories are then used to make recommendations.

From this we move on to methods of recommendation using unstructured data. In particular, in the online dating world, research has indicated that images are the most useful predictor of preference, and recommender systems based on user images have indeed proven to be very effective, in some cases even outperforming collaborative filtering methods.

4.2 Categorical Data-Based Reciprocal Recommendation

4.2.1 Data

Content-based recommender systems using categorical data are both the simplest form of recommender system to implement, and the earliest RRSs described in the literature, and therefore provide a good starting point for exploring their design. Categorical data itself on online dating services takes on a number of forms. Most commonly, users are required to display their age and location. Users intuitively look for other users around their own age and within meeting distance, and this is obvious enough that the vast majority of dating services have intuitively built them into recommendations since the 1990s.

Other information provided on user profiles depends on the service, and users may be able to choose which fields to fill in. It is common to include biometric categories such as height and weight, as well as information on hobbies, family information such as past marital status and children, as well as professional information such as job and income.

It is useful to formalize this. We let $\mathcal{A}$ be the set of all categorical attributes (such as age, location etc.), and any one attribute from that set is $a \in \mathcal{A}$. For an attribute a, let $\mathcal{V}_a$ be the set of possible values that attribute could take. For example, in the case of marital status this may be the set {single, married, divorced, widowed}. Then $v \in \mathcal{V}_a$ is a specific value of attribute a.

Modern dating platforms allow users to express interest in binary ways, often called sending a Like. This concept is central to many reciprocal recommender systems and is discussed in Section 3.2. While other interactions, such as exchanged messages, may provide more nuanced information about how successful a recommendation was, they are also much harder to usefully quantify, and users often move interactions off the service and onto private messages and phone calls before a serious relationship develops.

Analysing the relationship between Attributes and Likes can provide interesting information about the service, which can inform simple recommendations even without a full recommender system. For instance, it is often useful to know which attribute values are particularly attractive. We can define the popularity of an attribute value as the ratio between the number of Likes and the number of users with that attribute value.

Formally, we define U as the total set of users, and U_v as the subset of users who have attribute a with value v. Let L_v be the number of Likes received by users with attribute value v. Then, the popularity Pop(v) of attribute value v is defined as:

$$\text{Pop}(v) = \frac{L_v}{|U_v|} \tag{4.1}$$

The results of this analysis should, however, be used with care. Recommending popular users may appear successful in the sense that it often increases the overall number of Likes on the service, but popular users tend to have a low response rate, so reciprocal preference (*Match*) rate is likely to suffer. In order to balance this, a more sophisticated recommender system is required. This section describes several alternatives.

4.2.2 RECON: A Categorical Data Content-Based Recommender System

One of the earliest mentions of an RRS in the literature is *RECON* (REciprocal CONtent-based recommender for online dating) (Pizzato et al. [11]), which uses categorical data to make recommendations, and was successful in offline evaluation at accurately recommending users' preferences.

4.2.2.1 Reciprocal Preference Calculation

The service that RECON was originally evaluated on used a message sent as a binary indicator of preference, but the results have been replicated on other services using Likes. RECON operates by creating a *preference profile* for a user based on their previous messages. The preference profile is a list of numbers which represents which attributes a user prefers. For example, if Alice commonly sends messages to users who are young, blonde and non-smokers, her preference profile will be updated to reflect this, and her recommendations would be in line with these preferences.

Adopting the notation introduced in the previous section, let $\mathcal{A}_x = \{v_{x,a} : a \in \mathcal{A}\}$ be the profile of user x. We then define the preferences $p_{x,a}$ of a user x for an attribute a with possible values $\mathcal{V}_a$ as:

$$p_{x,a} = \{(v, l_{x,v}) : v \in \mathcal{V}_a \text{ and } l_{x,v} > 0\} \tag{4.2}$$

where $l_{x,v}$ is the number of times user x has liked (or sent messages to) users with attribute value v.

The overall preference profile of x is then:

$$\mathcal{P}_x = \{p_{x,a} : a \in \mathcal{A}\} \tag{4.3}$$

With preferences and preference profiles, we can now calculate Alice's preference for Bob based on Alice's preference profile, and Bob's profile. Informally, for each attribute value in Bob's profile, we divide the number of instances of that attribute value in Alice's preference profile with the total number of Likes that Alice has sent. This gives us Alice's preference for that attribute. The preference of Alice for Bob is then the sum of the individual attribute preferences, divided by the total number of attributes in Bob's profile.

Let $\mathcal{A}_B = \{v_{B,a} : a \in \mathcal{A}\}$ represent Bob's profile, where $v_{B,a}$ is Bob's value for attribute a. Let $l_{A,v_{B,a}}$ be the number of times Alice has liked users with the same attribute value $v_{B,a}$ as Bob's value for attribute a. Let L_A be the total number of Likes that Alice has sent.

Alice's preference for Bob, denoted by $P_{A \to B}$, can be defined as follows:

$$P_{A \to B} = \frac{1}{|\mathcal{A}_B|} \sum_{a \in \mathcal{A}} \frac{l_{A,v_{B,a}}}{L_A} \tag{4.4}$$

An alternative implementation divides the sum by the total number of attributes rather than by the number of attributes in that specific user's profile. Equation 4.4 may have unintentional bias towards users who have only filled in certain sections of their profiles, because with fewer attributes there is a higher likelihood of matching all of another user's preference, whereas using the total number of attributes might bias the system against users with incomplete profiles.

$P_{A \to B}$ represents a unidirectional preference of Alice for Bob, but recommendations based on unidirectional preference often do not produce effective reciprocal recommendations. RECON uses the Harmonic Mean, discussed in Section 2.4, to aggregate the preferences of Alice for Bob and Bob for Alice into a single reciprocal preference score, which is used to rank recommendations:

$$P_{A \leftrightarrow B} = \frac{2 P_{A \to B} P_{B \to A}}{P_{A \to B} + P_{B \to A}} \tag{4.5}$$

4.2.2.2 Example Preference Calculation

An example makes this process more intuitive. We use a dating service with four imaginary users, Alice, Bob, Charlie and David, and three attributes, Age (20-25, 25-30, 30-35), Height (Tall, Average, Short) and Smoker (Yes, Sometimes, No). The profiles for the users are listed in Table 4.1.

Profile	U_{Alice}	U_{Bob}	U_{Charlie}	U_{David}
Age	20-25	25-30	20-25	30-35
Height	Average	Tall	Average	Short
Smoker	Sometimes	No	Sometimes	No

Table 4.1: Profiles for four example users

Our users have also used the site to send Likes to other users on the service (we assume for the purposes of this example that they have not yet expressed positive or negative preference towards each other). The number of Likes they have sent to users with each attribute $p_{x,a}$ is shown in Table 4.2.

Preferences	$\mathcal{P}_{\text{Alice}}$	$\mathcal{P}_{\text{Bob}}$	$\mathcal{P}_{\text{Charlie}}$	$\mathcal{P}_{\text{David}}$
$p_{x,\text{Age}}$	(20-25, 10)	(20-25, 5)	(20-25, 1)	(20-25, 3)
	(25-30, 7)	(25-30, 10)	(25-30, 8)	(25-30, 4)
	(30-35, 3)	(30-35, 5)	(30-35, 1)	(30-35, 2)
$p_{x,\text{Height}}$	(Tall, 10)	(Tall, 8)	(Tall, 0)	(Tall, 5)
	(Average, 9)	(Average, 7)	(Average, 0)	(Average, 3)
	(Short, 1)	(Short, 5)	(Short, 10)	(Short, 1)
$p_{x,\text{Smoker}}$	(Yes, 8)	(Yes, 6)	(Yes, 8)	(Yes, 1)
	(Sometimes, 7)	(Sometimes, 4)	(Sometimes, 2)	(Sometimes, 5)
	(No, 5)	(No, 10)	(No, 1)	(No, 3)

Table 4.2: Preference profiles for four example users

Table 4.3 shows the unidirectional preferences as calculated by Equation 4.4, based on the values in the first two tables. (Preferences of users for themselves can be computed, but are meaningless in the RRS domain and therefore omitted.) Finally, Table 4.4 shows reciprocal preferences, which is the harmonic mean of the two unidirectional values and is therefore symmetrical along the diagonal.

4.2.2.3 RECON Implementation

The length of the formalization also hides to some degree how simple RECON is to implement, which is a significant advantage over more sophisticated algorithms. The following code is an example of a simple implementation of RECON in under 20 lines of code.

Preference	$P_{x\to\text{Alice}}$	$P_{x\to\text{Bob}}$	$P_{x\to\text{Charlie}}$	$P_{x\to\text{David}}$
$P_{\text{Alice}\to y}$	X	0.58	0.37	0.19
$P_{\text{Bob}\to y}$	0.54	X	0.27	0.38
$P_{\text{Charlie}\to y}$	0.55	0.53	X	0.15
$P_{\text{David}\to y}$	0.42	0.53	0.28	X

Table 4.3: Unidirectional preferences between all pairs of users

Reciprocal Preference	Alice	Bob	Charlie	David
Alice	X	0.56	0.45	0.27
Bob		X	0.36	0.43
Charlie			X	0.18
David				X

Table 4.4: Reciprocal preferences between all pairs of users

```
def calculate_preference(user_profile, other_values):
    total_likes = sum(sum(values.values()) for values in
        user_profile.values())
    preference = 0
    for attribute, value in other_values.items():
        if value in user_profile[attribute]:
            preference += user_profile[attribute][value] /
                total_likes
    return preference / len(other_values)
```

First, we define our calculate_preference function, which computes $P_{A\to B}$ given a user's preference profile and another user's profile values. This is essentially an implementation of Equation 4.4, which returns the average preference of Alice for Bob's attribute values based on her historical preferences.

```
def harmonic_mean(a, b):
    return 2 * a * b / (a + b) if (a + b) != 0 else 0
```

This function calculates the harmonic mean, and is used to aggregate $P_{A\to B}$ and $P_{B\to A}$ into $P_{A\leftrightarrow B}$.

```
 1  def calculate_reciprocal_preference(A_profile, A_values,
        B_profile, B_values):
 2      # Calculate A's preference for B
 3      P_A_to_B = calculate_preference(A_profile, B_values)
 4
 5      # Calculate B's preference for A
 6      P_B_to_A = calculate_preference(B_profile, A_values)
 7
 8      # Calculate the reciprocal preference using the Harmonic Mean
 9      P_A_B = harmonic_mean(P_A_to_B, P_B_to_A)
10
11      return P_A_B
```

The function `calculate_reciprocal_preference` essentially combines the two functions described above, calculating two preference scores and then returning the harmonic mean of the two values as the output.

```
 1  # Define the preference profiles for Alice and Bob
 2  alice_profile = {
 3      'Age': {'20-25': 10, '25-30': 7, '30-35': 3},   # Alice's
            likes for each age
 4      'Height': {'Tall': 10, 'Average': 9, 'Short': 1},   # Alice's
            likes for each height
 5      'Smoker': {'Yes': 8, 'Sometimes': 7, 'No': 5}   # Alice's
            likes for each smoking status
 6  }
 7
 8  bob_profile = {
 9      'Age': {'20-25': 5, '25-30': 10, '30-35': 5},
10      'Height': {'Tall': 8, 'Average': 7, 'Short': 5},
11      'Smoker': {'Yes': 6, 'Sometimes': 4, 'No': 10}
12  }
13
14  # Define Bob's and Alice's actual attribute values
15  alice_values = {'Age': '20-25', 'Height': 'Average', 'Smoker': '
        Sometimes'}
16  bob_values = {'Age': '25-30', 'Height': 'Tall', 'Smoker': 'No'}
```

The data above is provided as an example of how the implementation might operate. We have Alice and Bob's profiles, and their historical preferences. We use these as inputs to the algorithm. Note that the step of converting profile data to the form displayed above is skipped, but is often more labour intensive than the implementation of the algorithm itself. Some choices made by users are multiple choice (e.g. whether or not the user has children) and therefore can be added to the preference profiles as is, but some profile elements which influence Likes are closer to continuous variables such as age, and decisions need to be made about whether pre-processing such as bucketing is useful.

```
 1  calculate_reciprocal_preference(alice_profile, alice_values,
        bob_profile, bob_values)
```

Finally, we call the preference calculation function using Alice and Bob's profiles and their preferences.

4.2.2.4 Strengths and Weaknesses of RECON

RECON's main strength lies in its simplicity of implementation. This and the fact that it is an older algorithm which has a track record of experimental results in the context of being used as a baseline for other algorithms makes it a useful algorithm for an initial implementation of an RRS. It can then be used as a baseline to measure the performance of other, more sophisticated algorithms.

RECON is also relatively efficient to implement, in the sense that calculations based on processing categorical data are generally quicker than unstructured data such as text or images, which often require a reduction to meaningful embeddings before they can be efficiently processed. Because the naive implementation compares every user with every other user on the service, for a service with n users and $|\mathcal{A}|$ attributes, the complexity of calculating RECON recommendations for every user grows with $O(n^2|\mathcal{A}|)$, which makes real-time calculations challenging on larger servies. However, this is common in the context of content-based algorithms, and there are a number of methods of solving this such as clustering users to reduce the number of comparisons.

The most intuitive weaknesses are common problems with collaborative filtering algorithms. For example, data must be strictly categorical, and bucketing continuous data such as age creates problems at the edges of buckets. A user who usually likes other users who are 20-25 might also be happy with someone who is 26, but that user may be excluded from their recommendations because they missed the 20-25 bucket. The system also treats all attributes as if they are equally important predictors of attraction, which is unlikely to be the case. Dating services often have dozens of fields that it is possible for users to fill out, and when a user has a short history, coincidental correlations are likely to muddy preference score calculations.

A less intuitive but perhaps more significant problem with RECON in online dating environments is that many users may not be using categorical data to make decisions. This is especially the case in modern services, where users are shown a photos and given an immediate opportunity to make a decision about whether to Like that user. Informal research made public by online dating services has also alluded to the idea that especially photos may be significantly more important than categories to users' decision-making processes. The rest of this section looks at methods for using other types of data to calculate preference relations.

4.3 Image and Text-based Recommender Systems

4.3.1 Data

Aside from the categorical data, which has been the most widely used data in online dating services (mainly because of simplicity of use), unstructured data is also available. The most common types of unstructured data are:

1. One or more photos, which may be required to abide by certain rules (such as not containing the faces of people besides the user, and not containing nudity) but is otherwise the user's choice.
2. A text description, which may be a profile for an online dating service, or a short description of careers and accomplishments on a recruitment site. This would usually be at least a paragraph, and may be significantly longer depending on the user and the service.
3. Messages exchanged by users. These are usually constrained by certain rules, often forbidding explicit content or advertising, but are otherwise up to the users.

As mentioned in the previous section, informal industry research suggests that photos are the most important aspect that users pay attention to when deciding whether or not to send Likes. This is also intuitively likely to be the case when examining the way that modern online dating services present information. Searches and recommendations will usually show first and foremost a list of thumbnail images, often with other key information such as ages and locations. Users only see detailed profiles after clicking a photo thumbnail. They are therefore implicitly encouraged to use photos as an initial filter for whom they will and won't send Likes to.

If photos are indeed the most important factor in determining whether or not Likes are exchanged (and therefore whether or not communication can commence), a recommender system based on photos would have a higher ceiling for accuracy than one based on categorical data. The catch is that, as unstructured data, photos are much more difficult to interpret, requiring more sophisticated models to make recommendations. In part due to the difficulty of accessing online dating service data, and the privacy and PR concerns with making these kinds of models available, there has been very little research done in general on AI models of popularity and attractiveness based on user photos.

There is currently no public research into using unstructured text data for recommendations. However, it would be surprising if unstructured text data was not a useful predictor of mutual attraction, given that before exchanging messages, free text descriptions are the primary method for users to express their personalities. We will therefore look at methods of extracting useful information from text data under the assumption that depending on the service, it is a useful predictor of preference.

Messages seem intuitively like they would be very useful for reciprocal recommendation, but in reality are surprisingly difficult to interpret. It is tricky to convert messages into useful positive and negative signals, and difficult to establish a target for successful or failed communication exchanges. Some users may move their communications to email or phone conversations very quickly, where we cannot see the conclusion of the conversation, and other users may stop replying to their matches for non-negative reasons (they were busy, or they started dating someone else). There has thus far been no conclusive research done on recommendation using message content.

This is a more difficult problem than typical image-based AI problems such as object detection and recognition: attractiveness is a very abstract concept, and it is not well established how people decide whether they find someone physically attractive. Some research has been done into ideal face proportions and ratios [2], but this is usually done on extracted faces in isolation, and is not relevant to online dating services, where users take photos from arbitrary angles on arbitrary backgrounds, where parts of the image besides the user's face are likely to contribute to attractiveness.

It is also important to differentiate between *general attractiveness*, which indicates attractiveness in the sense of popularity and general appeal, and *personal attractiveness* in the sense of Alice being attractive to Bob specifically, regardless of her attractiveness in general. The former is of less interest because it is only minimally helpful when making recommendations: as discussed, popular users are not usually good candidates for recommendation. The latter is much more useful but also a more challenging problem: the model must be able to perform *few-shot learning* in the sense of being able to ascertain a user's preferences based off a small number of Likes. If a model requires thousands of examples of a user's preferences to learn who is personally attractive to them, it is unlikely to be useful in a real-world setting. There has been some research into machine learning methods for general attractiveness [6] but very little into personal attractiveness outside of the method described later in this chapter.

Unstructured data such as text and photos can't be used in its raw form. We must therefore use machine learning techniques to condense unstructured data into a meaningful representation within the context in which it is being used. This is known as *feature extraction*, and the following sections outline a few methods of doing this.

4.3.2 Image Data Feature Extraction

Most feature extraction methods for images involve modern machine learning methods. In particular, *Convolutional Neural Networks* (CNNs), which were introduced in Section 2.3.1, are suitable for feature extraction from images. Standard user-item recommender systems will often have items which are classified and labelled. For example, a shopping recommender system will list items as part of a category tree, and will often have additional tags for things like the brand. Where people are the focus of the recommender system, we almost never have such convenient categories which relate directly to the appearance of the person in a way which is useful for recommendation. For general purpose feature extraction, we therefore often have to make do with *Unsupervised Learning* (training with no labels) or *Self-Supervised Learning* (generating labels from the data).

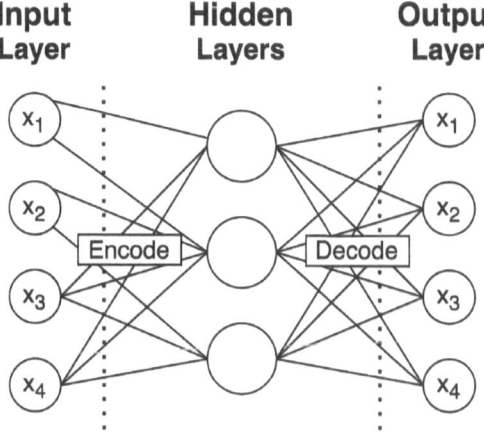

Fig. 4.1: Example Autoencoder

In either case, our first objective is to generate a vector which represents the image. Assuming we do not have labels, there are a number of options where unsupervised learning on images is concerned. A simple method is to use an *Autoencoder*, which is a neural network that consists of two parts, an encoder and a decoder, as depicted in Figure 4.1. The encoder takes an input x and learns to represent it as a one-dimensional vector z:

$$z = f_{enc}(x) \tag{4.6}$$

The decoder takes z and from it attempts to reconstruct x. We call the reconstruction $\hat{x}$:

$$\hat{x} = f_{dec}(x) \tag{4.7}$$

The encoder is often a standard CNN, and the decoder is commonly (although not necessarily) the mirror of the encoder (i.e. the same network, with the layers reversed). The network is trained with an image x as both the input and the output, so that the network learns to reconstruct $\hat{x}$ from z, and z can therefore be considered a meaningful lower-dimensional representation of x. This vector can then be used in various ways in recommendation. As a simple example, if Alice likes a number of users with photos $x_1, x_2, x_3, \ldots$ we can find their feature representations $z_1, z_2, z_3, \ldots$ from our pre-trained network, and then recommend a user with a similar photo, where similarity is based on the Euclidean distance between their feature recommendations:

$$D = ||z_m - z_n|| \tag{4.8}$$

More advanced methods of learning on unlabelled data have also recently been used in recommendation algorithms, where labels are generated before training. Positive training pairs are generated from an image and a slightly filtered version of the image (such as a blur or a contrast adjustment), and negative training pairs are two different images [4]. These have yet to be tested in reciprocal environments, but their usefulness in user-item recommendation implies that they would be effective in this context as well.

4.3.3 Text Data Feature Extraction

Similar to images, text data must also be converted to a simple, numerical format before we can use it for recommendations. Text data is not as dense as image data, however, and there are therefore a few simpler methods which do not rely on trained machine learning models. This section introduces one simple method which is easy to implement, and one machine learning-based method.

4.3.3.1 Term Frequency - Inverse Document Frequency

Term Frequency - Inverse Document Frequency (TF-IDF) is commonly used to find the most significant words in a document. It is easy to imagine how this might be useful in the context of reciprocal recommendation: users are often good matches for other users with whom they have things in common. For example, on a recruitment service, a company which mentions the language *Python* job listing might be a good match for a candidate who talks about their experience with *Python* on their CV.

We assume we have a corpus of documents; for example, all of the job listings on a recruitment service. TF-IDF starts by calculating the *Term Frequency* (TF) of every word in a given document. This is the number of times that a word is mentioned in a document, divided by the length of the document:

$$TF = \frac{\text{Number of times the term appears in the document}}{\text{Number of words in the document}} \tag{4.9}$$

This might give us some initial idea of significance, but many of these words are likely to be very common in general, such as "the" and "is". We use the *Inverse Document Frequency* (IDF) to reduce the weight of very common words:

$$IDF = \log\left(\frac{\text{Total number of documents}}{1 + \text{Number of documents containing the term}}\right) \tag{4.10}$$

The smaller the number of documents in the corpus containing a word, the more significant it is considered by the formula. The 1 is added to the denominator to ensure that the IDF does not reach zero, even when the word is in every document. The TF-IDF is then calculated as the product of the two values:

$$\text{TF-IDF} = \text{TF} \cdot \text{IDF} \tag{4.11}$$

By ranking terms in the document by TF-IDF, we can find the most significant terms and, by comparing these with the most significant terms of other users, can use this to make recommendations [1].

4.3.3.2 Machine Learning Text Methods

Similar to feature extraction from images, there has been a lot of progress in recent years in using more advanced machine learning models to extract features from text. In particular, *Attention*-based mechanisms have been particularly effective in extracting features from text. Attention is a mechanism whereby words in the sentence that are useful for helping to understand the meanings of other words are identified.

A thorough treatment of Attention is beyond the scope of this book, but we expand on it briefly here. Attention-based models are trained as neural networks. They consist of alternating *Attention Layers* and *Feed-Forward Networks*. The Attention Layers learn the relationships between words in a sentence, and the Feed-Forward Networks perform non-linear transformations on individual tokens. Attention-based models are used in a variety of tasks, including next word prediction. This makes them powerful as the basis for tasks such as developing chatbots and machine translation.

Moving past the details of implementation, the ability of these models to interpret meaning over sentences has made them useful tools in Recommender System design [12]. In particular, pre-trained models such as BERT which are available open-source can be fine-tuned on data from the environment in which the recommender system will operate, and used to output an embedding which represents a user's freetext profile. This embedding can then be used as a vector in the context of content-based recommendation in the same way as other numerical data, by applying an algorithm such as RECON.

4.3.4 ImRec: An Image-Based Recommender System

ImRec (Image Recommender System) (Neve et al. [9]) uses a machine learning model to determine personal attractiveness. At the time of writing, this is the only published algorithm which uses images to determine personal attractiveness, so it makes for an interesting case study. Specifically, the algorithm uses a subtype of CNN known as a *Siamese Network* to make recommendations.

4.3.4.1 Siamese Networks

A standard CNN accepts a single n-dimensional input, and calculates an output based on optimizing a number of hidden layers over a number of passes through a training dataset, through which it learns to recognise useful features. A Siamese Network, on the other hand, accepts two inputs and learns whether they represent members of the same class or members of a different class by reducing them to embeddings via a convolutional network, and then comparing the embeddings. Siamese Networks have been particularly successful in face recognition, where after training on a sufficient number of examples, they are able to accurately recognise new faces with only a few examples of a previously unseen face [8].

A Siamese Network is trained using pair inputs, usually of the form (x_a, x_p) and (x_a, x_n). In this case, x_a is the *anchor* image. x_p is an image from the same classification group as x_a. x_n is an image from a different classification group from x_a. For example, training a Siamese Network for face recognition, x_p might be a photo of the same person as the face x_a, and x_n might be a photo of a different person. The output label is defined as $Y \in \{0, 1\}$, where an output of 0 indicates that the the two inputs were of the same category, and 1 indicates the two inputs came from different categories. Thus, the network learns how to correctly classify photos of one person given an example.

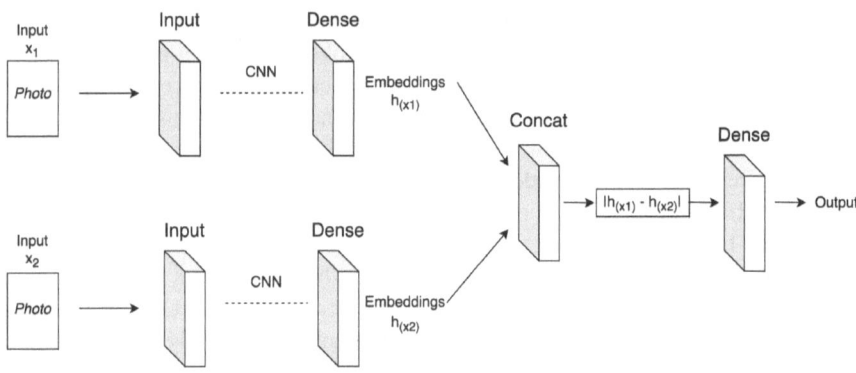

Fig. 4.2: An example Siamese Network

The structure of a Siamese Network is shown in Figure 4.2. The two identical CNNs are abbreviated for the sake of visibility, but the structure of the CNN would depend on the task, and would usually follow a standard structure with pooling and convolution layers, and a final k-dimensional dense layer where the output contains a set of embeddings that represent the input image. For input images x_1 and x_2, consider the output embedding vectors $f(x_1)$ and $f(x_2)$ respectively. The Euclidean distance between these two vectors is defined as:

$$D = ||f(x_1) - f(x_2)|| \tag{4.12}$$

A greater distance between the two vectors indicates a higher level of dissimilarity between the two inputs.

Siamese Networks are generally trained with a unique loss function known as *Contrastive loss*. Contrastive loss is a function in two parts:

1. Where $Y = 0$ i.e. the pair of inputs belongs to the same class, the loss function is $L_{similar} = \frac{1}{2}D^2$. This results in a small loss for short distances and a larger loss for larger distances.
2. Where $Y = 1$ i.e. the pair of inputs belongs to different classes, the loss function is $L_{dissimilar} = \frac{1}{2}\max(0, m - D)^2$. This function uses a threshold parameter called the *margin m*, which determines the size of loss for similar pairs. Where $D > m$, the distance is sufficiently large and so the loss is zero; when $D < m$ the loss increases quadratically with smaller distance. A smaller margin will result in a stricter classifier (i.e. higher precision, but lower recall).

Multiplying the similar function by $1 - Y$ and the dissimilar function by Y allows us to construct a single function for L, where the appropriate computation (for $L_{similar}$ or $L_{dissimilar}$) is used depending on the label:

$$L_{similar} = (1 - Y) \cdot \frac{1}{2}D^2 \tag{4.13}$$

$$L_{dissimilar} = Y \cdot \frac{1}{2}\max(0, m - D)^2 \tag{4.14}$$

So the final loss function is:

$$L = \frac{1}{2}((1 - Y) \cdot D^2 + Y \cdot \max(0, m - D)^2) \tag{4.15}$$

The Contrastive loss function in particular aims to increase separation between members of the same class, and members of different classes. It does this by increasing the Euclidean distance between members of different classes, and decreasing the distance between members of the same class.

4.3.4.2 RRS Network Design

Circling back to reciprocal recommendation, we can use a trained Siamese network to perform few-shot learning of user preferences. The data, described in the following section, uses Likes and Nopes to determine input and output data. This section first describes the structure of the network to be used.

In face recognition, $Y = 0$ when x_1 and x_2 are representations of the same face. In the RRS domain, $Y = 0$ is the case when both the inputs are photos of users for whom a given user indicated positive preference. $Y = 1$ is the case where the inputs are one photo of a positive preference expression and one photo of a negative preference expression from a given user. The system learns to differentiate between examples of positive and negative personal preference.

The CNN that represents the symmetrical part of the Siamese Network is visualised in Table 4.5. This is based on CNN used in the face recognition CNN *FaceNet*, under the assumption that faces are a significant part of attraction (based on experimental training results, this was the most effective structure of the networks tested).

Layer	Size-in	Size-out	Kernel	Param
input		100x100x3		0
conv1	100x100x3	100x100x3	7x7x3	444
maxpooling1	100x100x3	34x34x3	3x3	
normalization1	34x34x3	34x34x3		12
conv2	34x34x3	34x34x64	3x3x64	1792
maxpooling2	12x12x64	12x12x64	3x3	
normalization2	12x12x64	12x12x64		256
conv3	34x34x3	12x12x192	2x2x192	49344
maxpooling3	12x12x64	4x4x192	3x3	
conv4	4x4x192	4x4x384	2x2x384	295296
maxpooling4	4x4x384	2x2x384	3x3	
conv5	2x2x384	2x2x256	1x1x256	98560
conv6	2x2x256	2x2x256	3x3x256	590080
maxpooling5	2x2x256	1x1x256	3x3	
flatten	1x1x256	256		
dense1	256	256		65792
dense2	256	128		32896

Table 4.5: The structure of the CNN used as the symmetrical part of the network to create embeddings

It is possible that the optimal CNN structure depends on the RRS environment in which the algorithm is being used.

The final layers of the network follow the pattern outlined in Figure 4.2, with one difference. The output layer in standard tasks for a Siamese network usually a sigmoid layer for classification tasks. However, we are interested in ranking recommendations, so the output layer in RRS environments is a linear output node, which allows us to evaluate the output in the context of a number between 0 and 1.

Following the network design, it must then be trained on sufficient data to learn patterns of attraction in the context of the service. In ImRec's original evaluation, 500000 pairs of images were used to train the network. Depending on the service, more or less data may be required to obtain an effective result. Having trained the network, its usage is as follows: given an anchor image of Bob which Alice has shown preference for and a second image of Charlie for whom Alice's preference is unknown, the network outputs a value between 0 and 1 which represents how likely Alice is to like Charlie's photo given her positive preference for Bob's photo.

4.3.4.3 Recommendation Algorithm

The output of the Siamese Network detailed above produces a number between 0 and 1, which represents the preference of a user for a photo. While this is useful, it does not yet equate to a working RRS. This section explains how to use this result to make predictions about reciprocal preference.

The full algorithm is outlined in Figure 4.3. We can make predictions about Alice and Bob's preference for each other using previously liked photos. Step 1 in the

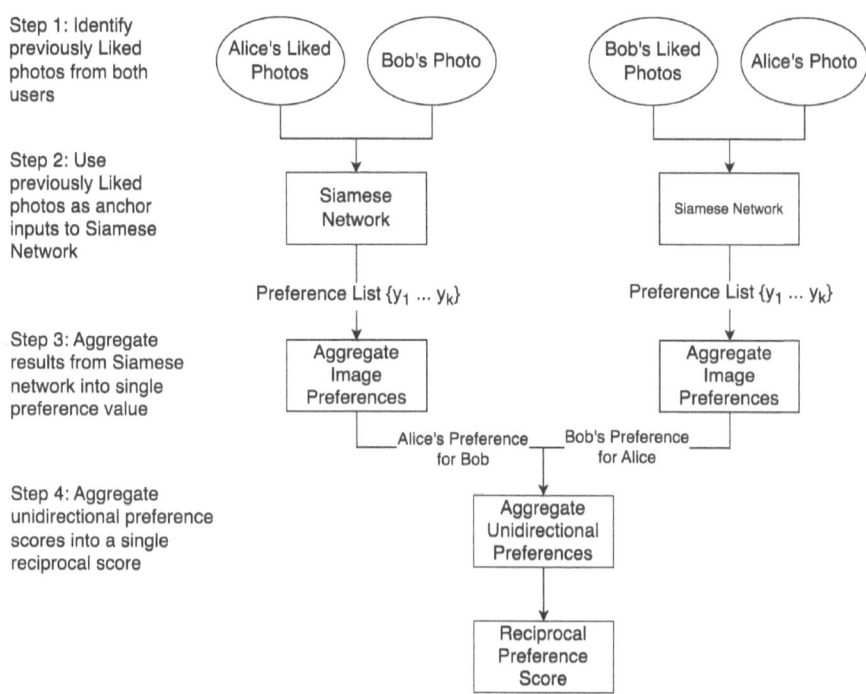

Step 1: Identify previously Liked photos from both users

Step 2: Use previously Liked photos as anchor inputs to Siamese Network

Step 3: Aggregate results from Siamese network into single preference value

Step 4: Aggregate unidirectional preference scores into a single reciprocal score

Fig. 4.3: ImRec Algorithm

diagram is to extract the most recent k users from Alice's preference history, and find their photos $x_1...x_k$. These photos form the anchor inputs x_a. Then, we use Bob's photo x_B to make input pairs $\{(x_1, x_B), ..., (x_k, x_B)\}$.

Step 2 is to evaluate these pairs using the trained Siamese Network. The output from this is a series of numbers between 0 and 1, $Y_{A \to B} = \{y_1, ..., y_k\}$. If we repeat this process for Alice's photo, using Bob's previous preferences as anchors, we can obtain a second list $Y_{B \to A} = \{y_1, ..., y_k\}$, which represents Bob's preference for Alice's photo given photos that he has previously expressed preference for.

As Step 3, the two lists of scores must be aggregated into two numbers, which represent the unidirectional preference of Alice for Bob $P_{A \to B}$ and the preference of Bob for Alice $P_{B \to A}$. During the design of ImRec, various aggregation methods were tried, including Pythagorean means and using the scores as inputs to a further machine learning model, and small differences in performance were observed. These differences are likely to depend on the nature of the service. For simplicity, we present the harmonic mean, which was effective in the original testing, and has a history of general effectiveness in RRS environments:

$$P_{A \to B} = \frac{k}{\sum_{i=1}^{k} \frac{1}{y_i}} \qquad (4.16)$$

The same equation using Bob's results produces his preference for Alice. Finally, as Step 4, we aggregate the two preferences into a single score which represents the reciprocal preference between Alice and Bob. For this, we also use the two-value harmonic mean, in the same way as we did with RECON:

$$P_{A \leftrightarrow B} = \frac{2P_{A \rightarrow B} P_{B \rightarrow A}}{P_{A \rightarrow B} + P_{B \rightarrow A}} \tag{4.17}$$

This results in a single reciprocal preference score which represents the preference of Alice and Bob for each other.

4.4 Explaining Recommendations

Explainability of a recommender system refers to how easy it is for the user to understand exactly why they received a particular recommendation. In user-item recommender systems on popular services such as *Netflix*, we are often presented with a reason why a particular product has been recommended to us. For example, 'Recommended to you because you watched *The Godfather*'.

Explanations for why recommendations have been chosen increases the chance that users will click on those recommendations [5]. Using a movie streaming service as an example, if we simply present a list of titles to the user, they may not immediately understand why they are supposed to be appealing. If they are told that a certain movie has been recommended to them 'because of your interest in Horror' or 'because you watched other movies with Brad Pitt', they immediately get some information about why they might like it. It therefore improves *trust* in the recommender system: it is no longer a black box which is analysing the user with some unknown method, but recommending in the same way that a friend might recommend considering the user's preferences.

Content-based filtering offers a unique advantage in explainability. Unlike collaborative filtering, it bases suggestions on specific profile elements, allowing for tailored explanations. An algorithm such as RECON, which makes recommendations based on categorical data is particularly well suited for this, as we can tailor human-readable sentences for recommendations from each profile category. Recommender systems using unstructured data such as freetext and images are a little more difficult to explain, as they often come from machine learning models from which appropriate human-readable explanations are a little more difficult to extract.

The key equation in RECON's preference calculation which will help us here is Equation 4.4 [7]. Recall that RECON calculates Alice's preference for Bob $P_{A \rightarrow B}$ from the average preference of Alice for each of Bob's attributes based on her historical data. We extract the term which denotes Alice's preference for one of Bob's attributes a:

$$\frac{l_{A,v_{B,a}}}{L_A} \tag{4.18}$$

Recall from Section 4.2.2 that this is the number of times Alice has Liked someone with Bob's attribute value a divided by the number of total Likes Alice has sent. Then, we base our explanation on the maximum value of this attribute preference:

$$\max_{a \in A} \frac{l_{A,v_{B,a}}}{L_A} \tag{4.19}$$

We can then use the attribute value a to generate explanations. For example, if Alice has mostly Liked men who want children, and Bob has indicated that he wants children, we can explain Bob's recommendation as, 'Recommended to you because your history indicates that you prefer people who want children!'. Instead of using the max, we can also use the top k attributes and list all of them as reasons.

This kind of explanation based on categorical preferences is common in user-item recommendation. In reciprocal environments, our recommender system calculates not only Alice's preference for Bob, but also Bob's preference for Alice. Research indicates that there is benefit to RRSs not only explaining to Alice why she might like Bob, but also why Bob might like her back. This makes intuitive sense: online dating services often make Likes a limited resource to discourage users from using them indiscriminately. Users are therefore more likely to Like another user if they feel that they have a good chance of being Liked back and matching with that user.

Generating specific reciprocal explanations is done in the same way as generating one-way explanations - using the mirror of Equation 4.19 to find Bob's preferences for Alice's attributes. We can then explain both sides of the reciprocal preference equation: 'Bob was recommended to you because he also wants children. We also think that Bob is likely to respond positively to you, because he prefers people who have cats!'.

Interestingly, however, while research is clear about the benefits of providing one-way explanations, it is inconclusive about the effects of providing specific reciprocal explanations [7]. The primary research paper on this topic indicates that less specific explanations (e.g. '... and Bob is also likely to respond positively to you') may have a better effect. Psychologically speaking, it is not clear why this might be, and more research is needed to determine the best way of providing reciprocal explanations.

4.5 Choosing Between Content-Based and Collaborative Filtering

We have now looked at examples of content-based and collaborative filtering RRSs. We finish this chapter with a brief discussion of how to choose between them.

Where the choice between content-based and collaborative filtering algorithms is concerned, it is worth being aware that we do not always have to make this choice. Hybrid algorithms, discussed in Chapter 5, combine more than one algorithm to harness the strengths of both. Some of the most successful algorithms have been hybrid filtering systems which combine a content-based filtering algorithm and a collaborative filtering algorithm.

However, it is still worth being aware of the strengths of both. In general, collaborative filtering algorithms outperform content-based filtering algorithms. This has been true since the advent of recommender systems, where both have been used on the same dataset. It is easier to be sure about this in the case of user-item recommender systems, where large, public datasets are available on which sophisticated variants of both systems have been tested. It is harder to be sure about this in reciprocal environments, because for privacy reasons, large public datasets containing user preferences do not exist, and research papers are performed on private datasets of differing sizes. However, the evidence so far indicates that this is the case for RRSs too.

The one exception to this performance is in *cold-start situations*. This is the problem of making recommendations for a user new to the service with very little preference history, and is a vital part of recommendation. New users are much more likely to immediately leave the service if they don't find what they're looking for, and it is therefore essential to be able to make good recommendations for them. Straightforward collaborative filtering algorithms tend to be slightly weaker in this situation. Intuitively, meaningful correlations with other users are usually not available after only a few expressions of preference. In this situation, recommending users with similar profiles might not be a bad strategy.

References

1. Aggarwal, C.C., et al.: Recommender systems, vol. 1. Springer (2016)
2. Bashour, M.: An objective system for measuring facial attractiveness. Plastic and reconstructive surgery **118**(3), 757–774 (2006)
3. Bell, R.M., Koren, Y.: Lessons from the netflix prize challenge. Acm Sigkdd Explorations Newsletter **9**(2), 75–79 (2007)
4. Deldjoo, Y., Schedl, M., Cremonesi, P., Pasi, G.: Recommender systems leveraging multimedia content. ACM Computing Surveys (CSUR) **53**(5), 1–38 (2020)
5. Herlocker, J.L., Konstan, J.A., Riedl, J.: Explaining collaborative filtering recommendations. In: Proceedings of the 2000 ACM conference on Computer supported cooperative work, pp. 241–250 (2000)
6. Kagian, A., Dror, G., Leyvand, T., Meilijson, I., Cohen-Or, D., Ruppin, E.: A machine learning predictor of facial attractiveness revealing human-like psychophysical biases. Vision research **48**(2), 235–243 (2008)
7. Kleinerman, A., Rosenfeld, A., Kraus, S.: Providing explanations for recommendations in reciprocal environments. In: Proceedings of the 12th ACM conference on recommender systems, pp. 22–30 (2018)
8. Koch, G., Zemel, R., Salakhutdinov, R., et al.: Siamese neural networks for one-shot image recognition. In: ICML deep learning workshop, vol. 2, pp. 1–30. Lille (2015)
9. Neve, J., McConville, R.: Imrec: Learning reciprocal preferences using images. In: Proceedings of the 14th ACM Conference on Recommender Systems, pp. 170–179 (2020)
10. Neve, J., Palomares, I.: Latent factor models and aggregation operators for collaborative filtering in reciprocal recommender systems. In: Proceedings of the 13th ACM conference on recommender systems, pp. 219–227 (2019)
11. Pizzato, L., Rej, T., Chung, T., Koprinska, I., Kay, J.: Recon: a reciprocal recommender for online dating. In: Proceedings of the fourth ACM conference on Recommender systems, pp. 207–214 (2010)

12. Sun, F., Liu, J., Wu, J., Pei, C., Lin, X., Ou, W., Jiang, P.: Bert4rec: Sequential recommendation with bidirectional encoder representations from transformer. In: Proceedings of the 28th ACM international conference on information and knowledge management, pp. 1441–1450 (2019)
13. Volkovs, M., Yu, G.W., Poutanen, T.: Content-based neighbor models for cold start in recommender systems. interactions **322**, 92–949 (2017)

Chapter 5
Hybrid Filtering and Additional Approaches

5.1 Introduction

Chapter 3 and Chapter 4 described the two types of recommender systems which are primarily used in reciprocal environments, collaborative filtering and content-based filtering. We also examined some subdivisions between types of recommender systems - for example, model-based and memory-based collaborative filtering systems.

While these systems are often effective by themselves, they can be combined into *Hybrid Recommender Systems*. Hybrid systems attempt to improve on using a single type of recommender system by capturing the advantages of several different types [5]. For example, a collaborative filtering algorithm based on latent factor models might be effective when we have a certain amount of data for a user, but provide weaker recommendations for a user who has only Liked a few other users. In this case, we might try using a content-based filtering algorithm to address this weakness for users with fewer data points.

Hybrid systems have traditionally been the most effective recommender systems where evaluation metrics are concerned. The systems that eventually performed the best during the *Netflix Prize Challenge* were largely hybrid systems, which combined content-based filtering methods with collaborative filtering [2]. In spite of this, there has been relatively little research into hybrid systems in reciprocal environments.

The main reason for this is that hybrid systems are naturally more complicated to design than a single recommender system. Not only do two or more separate systems need to be developed and modified for the problem domain, but decisions need to be made about how to combine them into a useful hybrid. They are also potentially less efficient, as the most complex system in the hybrid will always be the bottleneck. As a result, we usually turn to hybrid systems once all possible performance optimizations have been squeezed out of a single recommendation model. In the case of reciprocal recommendation, there are still a great many potential RRS systems that have yet to be tested due to lack of access of researchers to large datasets.

This is not to say that hybrid systems do not exist. Several hybrid RRSs have been proposed which attempt to improve on the results of classic algorithms such as RCF

J. Neve, *Reciprocal Recommender Systems*, SpringerBriefs in Computer Science, https://doi.org/10.1007/978-3-031-85103-2_5

(Section 3.3.1) and LFRR (Section 3.4.2). However, this is a field still in relative infancy. The algorithms in this chapter are presented as examples of how hybrid systems might be used to overcome specific problems, but implementing them out-of-the-box may not be optimal. Where a hybrid approach is desirable, developers should not shy away from experimenting with alternatives that may be more suitable for their domain and the problem they are trying to solve.

This chapter first discusses the different types of hybrid filtering in a general sense, and the contexts in which they are used. These categories of hybrid filtering mirror those found in user-item recommender systems, and form a necessary basis for understanding the design of specific algorithms.

We then give two examples of hybrid recommender systems which have been effective in reciprocal environments specifically, and how to implement them. Firstly, *CCR* represents a relatively straightforward combination of content-based and collaborative filtering to improve on pure collaborative filtering. Then, we introduce *RWS*, an algorithm which uses hybridization to tackle one of the problems that plagues recommender system developers: that of popular users appearing in everyone's recommendation lists.

We end this chapter with a discussion of the costs and benefits of implementing hybrid system in RRS environments. There are a number of interesting trade-offs involved in this decision: hybrid systems can be more effective and, in user-item recommender systems, often outperform content-based or collaborative filtering by themselves. However, there are some additional considerations where RRSs are concerned.

5.2 Types of Hybrid Systems

Hybrid recommender systems are a multifaceted field, and there are a great many different ways of combining two or more recommender systems in order to improve on the results of the individual system. This section describes a the most common of these methods, and then gives two examples of these methods applied to reciprocal recommendation [4].

It is worth being aware that while this book describes four methods of creating a hybrid recommender system, there are a number of alternative and less common methods. However, given the sparsity of hybrid reciprocal recommenders, and the lack of evidence about whether they are significantly more effective than more straightforward methods, we stick to the four most common ones for the purposes of this chapter.

Figure 5.1 shows the four types of hybrid system discussed below in diagram form.

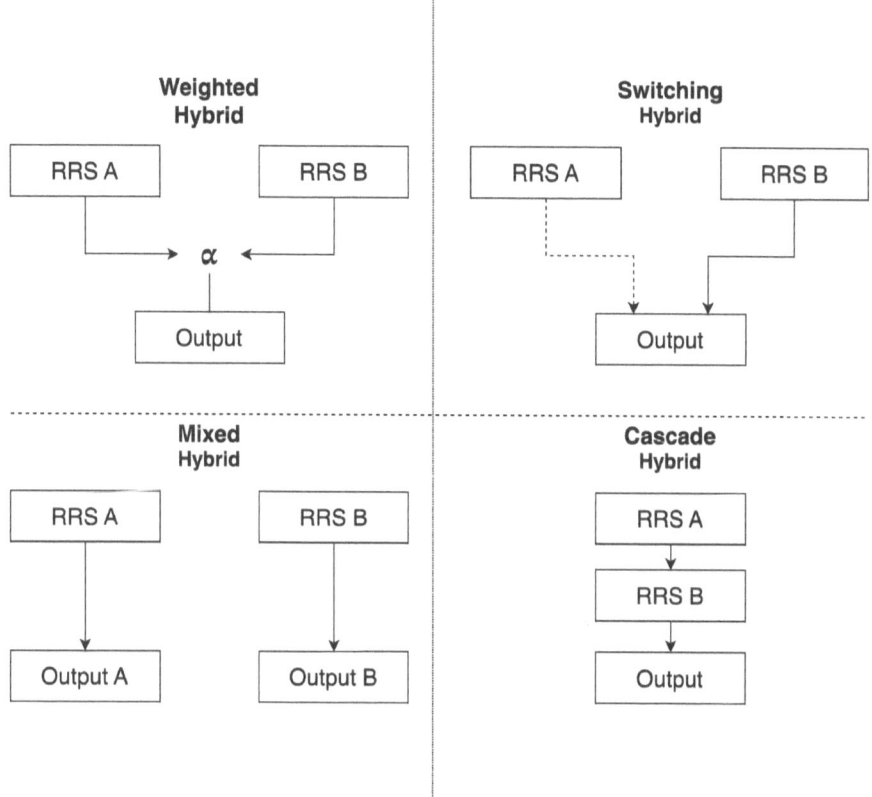

Fig. 5.1: Main types of Hybrid RRS

5.2.1 Weighted Hybrid

A weighted hybrid system combines the results of multiple different recommender systems into a single score which represents the reciprocal preference of Alice and Bob $P_{A\leftrightarrow B}$. A weighted hybrid system might use a simple linear combination, with a parameter α which controls the weight given to each of the systems. For example, if we have two reciprocal recommender systems which we want to make into a hybrid, with $P_{A\leftrightarrow B}^{(1)}$ as the reciprocal preference calculated by the first system and $P_{A\leftrightarrow B}^{(2)}$ as the preference calculated by the second system. Then we can calculate our overall reciprocal preference by:

$$P_{A\leftrightarrow B} = \alpha \cdot P_{A\leftrightarrow B}^{(1)} + (1 - \alpha) \cdot P_{A\leftrightarrow B}^{(2)} \tag{5.1}$$

In some RRS environments, we know that our preference equation is intuitively unbalanced. For example, in recruitment, candidates often choose how they apply for jobs in a fundamentally different way from how companies choose candidates.

The two preferences might also have different values as parts of the final calculation. For example, a prestigious company can often pick and choose its candidates. It therefore might not make sense to use the same recommender system to calculate both sides of the preference equation. In this situation, we can calculate $P_{A \to B}$ and $P_{B \to A}$ using different systems, and take the weighted average to give us a reciprocal preference:

$$P_{A \leftrightarrow B} = \alpha \cdot P_{A \to B} + (1 - \alpha) \cdot P_{B \to A} \qquad (5.2)$$

A weighted hybrid system might have *static weights* or *dynamic weights*. Static weights assume the same balance of different types of recommender system across all users. For example, a static weighted hybrid system might use 30% content-based filtering, 70% collaborative filtering for everyone. A dynamic weighted hybrid system would adjust the weights depending on the user. For example, under the assumption that collaborative filtering systems suffer more from the cold-start problem, in the context of a hybrid system consisting of a collaborative filtering system and a content-based filtering system, we might increase the weight of the collaborative filtering as the user accumulates preference data over time.

5.2.2 Switching Hybrid

A switching hybrid system utilises two or more recommender systems, and changes the system used depending on the context. We might switch the system depending on some specific context. For example, using content-based recommendations initially, and then collaborative filtering recommendations after the user has accumulated a certain amount of preference data helps to avoid the cold-start problem. We might also switch based on domain knowledge of our users. For example, in an online dating service, if we find that younger users are more likely to base decisions on appearance, it might be effective to use an image-driven content-based recommender for users under a certain age.

It is worth being aware that some recommender systems in the user-item domain calculate a confidence value for their recommendations, and switch based on which system can make recommendations with the greatest confidence. This has yet to be applied in reciprocal environments, however.

Switching systems can be challenging to design, because the switching criteria is often not straightforward, and a set of criteria that would improve the recommendations of some users might disadvantage other users. Determining a set of evidence-based switching criteria that are not arbitrary introduces significant additional complexity.

5.2.3 Mixed Hybrid

A mixed hybrid system aims to display recommendations from more than one recommender system simultaneously. This is not always practical - some services do not benefit from displaying very large numbers of varied recommendations to a user. However, where it is practical, it can have some advantages, as different systems which produce recommendations based on different types of logic can increase the variety of options displayed to a user.

While mixed hybrid systems allow us to display a wider variety of recommendation options without having to implement any particular logic in order to balance multiple systems or decide between them, they can have some disadvantages. Depending on the service, users might feel like they are being bombarded with recommendations if they have multiple sets coming from different sources. It can also increase the burden of *explainability* - users might have questions about why they have multiple lists of recommendations which appear different to each other.

5.2.4 Cascading Hybrid

A cascading hybrid system uses the results of one recommender system to refine the results of another system. In this model, one system is usually used to create a longer preliminary list of potential recommendations. The second system will usually then be used to refine this list, either by removing and adding items to it or by reordering it.

Cascading algorithms have the advantage of being relatively simple to implement, and are often more efficient than other models. Since the second algorithm in the cascade will be operating on a vastly reduced dataset, we usually only need to worry about the complexity of the first algorithm, and not about how to run two algorithms concurrently on large-scale data. They are also usually easy to evaluate, as we can directly compare the performance of the first algorithm with the performance of the cascade. The main disadvantage is that the second system in the cascade is less able to overcome the disadvantages of the first. For example, if a collaborative filtering algorithm which performs poorly in cold-start situations is used as the first system, the second system may end up refining a set of sub-optimal recommendations.

5.3 CCR: A Cascading Hybrid Reciprocal Recommender

Content-Collaborative Reciprocal Recommender (CCR) (Akehurst et al. [1]) is an algorithm which uses a combination of content-based and collaborative filtering to attempt to improve on the results of either system, and to mitigate the cold-start problem, which is a common objective of hybrid systems in user-item recommendation. It benefits from having a clear and easy-to-understand construction, and being relatively simple to replicate and test. It also deals with reciprocity in a unique way which differentiates it slightly from the algorithms discussed in the previous chapters, which use an aggregation function to combine the results of two symmetrical preference predictions.

The philosophy behind CCR is that similar users by profile are also similar in who they Match with. The CCR algorithm consists of four steps:

1. Calculate a distance metric between a target user, Alice, and all other users on the service $(X_1, X_2, \ldots, X_n)$. Then use this distance metric to find the closest (i.e. most similar) k users to Alice.
2. For each user who is similar to Alice, find the set of users with whom they have Matched $(M_{X_i} = \{Y_1, Y_2, \ldots, Y_m\})$.
3. For each user who has matched with a similar user to Alice, calculate a *Support* score, which is equal to the number of times they appear in the sets computed in the previous step.
4. Rank the resulting users by descending Support score to established a ranked list of recommendations.

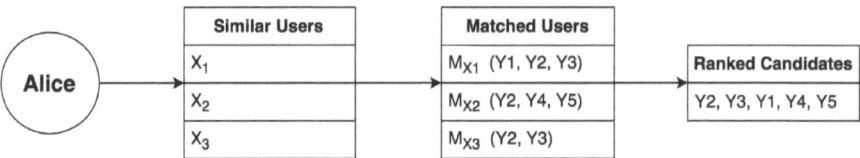

Fig. 5.2: A visual representation of the CCR algorithm

Figure 5.2 is a visual representation of CCR, inspired by Figure 3 in the paper which introduced the algorithm [1]. It shows the content-based system for calculating similar users, and then the collaborative filtering system which refines the list based on the users Matched with those similar users, finally giving us a list of recommendation candidates. This figure is inspired by Figure 3 in the paper where CCR was originally described.

We now explain each of these steps in a little more detail. Firstly, a distance metric $D(X, Y)$ is calculated between two users' profiles as the sum of the distances between individual attributes $d(v_{X,a}, v_{Y,a})$ where $v_{X,a}$ is the value of attribute a of user X. The original algorithm separates continuous and discrete variables, and encodes them differently.

For a continuous attribute a such as *Age* or *Height*, the distance between the value of Alice's attribute $v_{A,a}$ and Bob's attribute $v_{B,a}$ is as follows:

$$d(v_{A,a}, v_{B,a}) = \begin{cases} 0 \text{ if } |v_{A,a} - v_{B,a}| \leq 5 \\ 1 \text{ if } 5 < |v_{A,a} - v_{B,a}| \leq 10 \\ 2 \text{ if } |v_{A,a} - v_{B,a}| > 10 \end{cases} \tag{5.3}$$

Discrete variables with selectable categories such as *Body Type* and *Smoker* are treated differently. Each discrete variable is first converted to *Gray Code* [6], which is a binary form where the representations of consecutive categories differ only by one bit. (The specifics of the conversion are a little long winded, and therefore outside the scope of this book, but not complex.) For example, if *Body Type* has three possibilities, {*Slim, Average, Plus-Size*}, then they would be encoded as {(*Slim*, 00), (*Average*, 01), (*Plus-Size*, 11)}.

The distance $d(v_{A,a}, v_{B,a})$ is then the *Hamming Distance* between two variables in Gray Code. The Hamming Distance between two variables $d_H(x, y)$, where x and y are equal length binary strings $(x_0, x_1, \ldots, x_n)$ and $(y_0, y_1, \ldots, y_n)$ is equal to the number of bits in the strings which are different:

$$d_H(x, y) = \sum_{i=0}^{n} (x_i \neq y_i) \tag{5.4}$$

Using the Gray Code representation, two attribute values from consecutive categories will have a smaller Hamming Distance. Following on from the previous example, *Slim* (00) and *Average* (01) only have a Hamming Distance of 1, but *Slim* and *Plus-Size* (11) have a Hamming Distance of 2.

Now we have a distance function for both continuous and discrete variables, we can define a distance function for two user profiles. The distance between Alice with profile $\mathcal{P}_A$ and Bob with profile $\mathcal{P}_B$ over all attributes $\mathcal{A}$ is:

$$D(A, B) = \sum_{a \in \mathcal{A}} d(v_{A,a}, v_{B,a}) \tag{5.5}$$

This is the distance metric with which the original implementation of CCR was designed and tested, and therefore represents a suitable example metric. However, this is not necessarily suitable for all environments. In implementing CCR, algorithm designers should not feel constrained by this specific metric, but should design a distance metric which represents as accurately as possible the content-based similarity between two users.

In order to calculate Alice's recommendation list, we first calculate the distance metric between her and other users in the same category who can Match with the same users (e.g. if Alice is a woman who likes men, the distance metric would be calculated between her and other women who like men). We then take the k users closest to Alice and for each of them, find the sets of users with whom they have Matched. So if $M(X, Y)$ is a Match between X and Y, and M_X is the subset of users U with whom X has Matched:

$$M_X = \{Y \in U : (X, Y)\} \tag{5.6}$$

Having determined sets of candidate recommendations using content-based methods, we move on to refining and ranking this list with collaborative filtering. For a set of k users close to Alice $(X_0, \ldots, X_k)$, we find $(M_{X_0}, \ldots, M_{X_k})$. It is expected that some of these sets of M_{X_i} will have elements in common with each other i.e. some similar users to Alice will have matched with the same users as each other. We finally define a Support function $\text{Sup}(Y)$, which is the number of times a user Y occurs in the sets of Matched users. To formalize this, we define an indicator function:

$$1_{M_{X_i}}(Y) = \begin{cases} 1 \text{ if } Y \in M_{X_i} \\ 0 \text{ if } Y \notin M_{X_i} \end{cases} \tag{5.7}$$

And then sum over the sets of Matched users:

$$\text{Sup}(Y) = \sum_{i=0}^{k} 1_{M_{X_i}}(Y) \tag{5.8}$$

We can then rank recommendations by Support. If Bob has a high Support, it means that a number of users similar to Alice have Matched with Bob, and he would therefore be likely to have strong reciprocal preference with Alice.

The designers of CCR indicated that it was particularly adept at solving the Cold-Start Problem: where new users to a system aren't presented with effective recommendations by collaborative filtering algorithms until they have built up a preference history. This is a significant problem, because services often lose a large number of new users if they feel that they cannot find what they are looking for. New users by definition have very little invested, and they are therefore quick to move to other services. CCR calculates similarity based on a content-based distance metric instead of preference history, which means that it can give recommendations as soon as the user has filled in their profile. As the user's preference history expands, the collaborative filtering-based Support score ranks and refines these recommendations.

5.4 RWS: A Weighted Hybrid System for Balancing Popularity

Standard user-item recommender systems which use collaborative filtering often suffer from a bias towards very popular items, especially if they use binary data such as purchases on a shopping site as a metric for success. An item which many users buy is likely to be recommended as a result of many users having it in common with each other. As a result of being recommended, it is purchased more often, which leads to it being recommended to more and more users. This is known as a *feedback loop*.

It is a problem for popular items to get recommended so often that lesser known items are never seen by users, and some user-item recommender systems try to enforce *fairness* (so that all items are recommended at least occasionally), from the user's point of view it can be an advantage. Popular items are, after all, often popular for a reason, and buying an item with thousands of positive reviews often provides a level of assurance in the product (and therefore the shopping service) that three positive reviews do not.

However, RRS environments suffer much more from this problem. Popular users on online dating services are often inundated with so many messages and contacts that they are only able to respond to a very small percentage of them. If Alice is very popular and we recommend her to Bob, Bob might well send a Like (popular users are popular for a reason: they tend to be universally appealing) but there is only a small chance he will get a response. If this happens repeatedly, he may start to feel negatively about the service. On the other side of the equation, Alice may also start to feel overwhelmed by getting a large number of Likes every day that she is not able to respond to.

In theory, the preference aggregation method should combat this to some extent. If Alice is popular, for a given user x, $P_{x \to \text{Alice}}$ is more likely to be a high value, because many users Like Alice. $P_{\text{Alice} \to x}$ is likely to be lower, because Alice only responds to a small proportion of the Likes she receives. Using the harmonic mean for aggregation will often mean that two scores which are close together will be aggregated to a higher result than a lower value with a higher value, so pairs which include Alice are likely to be aggregated to lower scores. However, this is not infallible. Especially in algorithms like RCF, where scores tend to cluster together towards 0 in any case, a user with a sufficiently high score can significantly skew the result so that they are often recommended.

Section 3.3.1 introduced *RCF* [9], a collaborative filtering algorithm based on kNN style logic, which was shown to be effective in online dating situations. This section describes an example of a weighted hybrid reciprocal recommender system, *Reciprocal Weighted Score* (RWS) (Kleinerman et al. [7]). RWS uses a memory-based collaborative filtering algorithm to predict preference $P_{A \to B}$, and a model-based content-based filtering algorithm to predict reciprocal preference $PR_{B \to A}$ and then combines them using a weighted parameter α_A:

$$RWS_{A \leftrightarrow B}(\alpha_A) = \alpha_A \cdot P_{A \to B} + (1 - \alpha_A) \cdot PR_{B \to A} \qquad (5.9)$$

The parameter α_A is adjusted dynamically depending on A's preference history, where A is the user receiving the recommendation. In this section, we outline the general construction of RWS, and how the inputs and weights to Equation 5.9 are calculated.

RWS is an extension to RCF which conjectures that the two sides of the preference equation (the user receiving the recommendations and the user being recommended) represent the combined solutions of two fundamentally different problems. If Bob is a recommendation candidate for Alice, we should calculate Alice's preference for Bob to predict whether she will Like him. However, in Bob's case, we want to know not what his preference for Alice is, but whether or not he would respond positively to a Like from Alice.

This is quite a nuanced difference. Bob's preference for Alice is a factor in whether or not he replies, but so are lots of other things: how many Likes Bob gets in general (popular users are less likely to respond); how picky Bob is about responding to Likes; how often Bob logs into the service and so on. RWS aims to include these factors in its reciprocal preference calculation.

The decision of whether or not to recommend Bob to Alice depends on two scores. $P_{A \to B}$, as before, is the preference of Alice for Bob. This is usually calculated by RCF, although it could potentially be calculated by any collaborative filtering algorithm. $PR_{B\to A}$ is the probability that Bob will reply to Alice. In steps, the algorithm is as follows:

1. In choosing recommendations for Alice, first calculate $P_{A\to B}$, the preference of Alice for Bob.
2. Next, calculate $PR_{B\to A}$, the probability that Bob will reply to Alice.
3. Finally, combine these two scores into $RWS_{A\leftrightarrow B}(\alpha_A)$ using a weighted average, with parameter α_A.

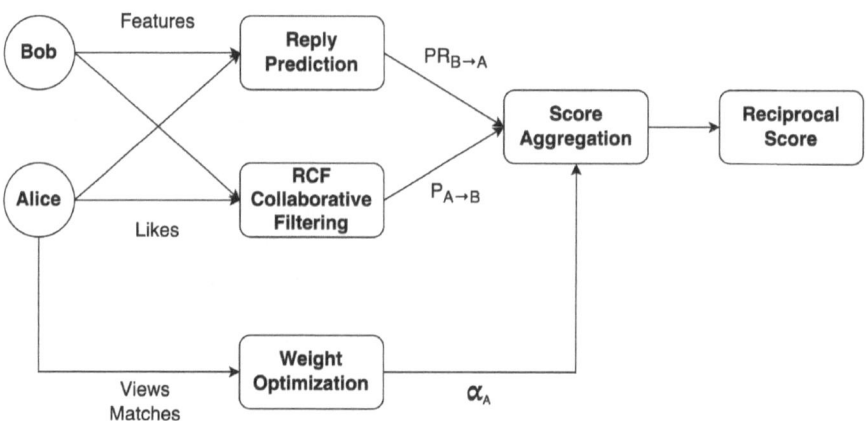

Fig. 5.3: A visual representation of the RWS algorithm

Figure 5.3, inspired by Figure 2 in the original paper [7], is an overview of RWS, including all the steps described above. Note the added complexity over and above implementing RCF: we need to calculate the collaborative filtering preference $P_{A\to B}$, the probability of reply $PR_{B\to A}$ and optimize the weight α_A in order to calculate a reciprocal score. These steps are explained in detail below.

$PR_{B\to A}$ is calculated using a machine learning model trained on features of both A and B. These are general pieces of information about the two users extracted from the data, which predict how likely Bob is to reply to Alice by assessing how likely Bob is to reply in general, and how likely Alice is to get replies. Some example features are as follows:

- Number of Likes Bob has received.
- Percentage of times Bob responded positively to being sent a Like.
- Number of times Bob logged into the service during the last week.
- Number of Likes Alice has received.
- Number of times Alice received a positive response to a Like.

Having extracted k relevant features, we build these into a feature vector:

$$f = (f_1, f_2, f_3, \ldots, f_k) \tag{5.10}$$

These features are then used to fit a non-parametric machine learning model, commonly *AdaBoost* [8]. While a thorough treatment of machine learning model types is outside the scope of this book, unfamiliar readers can consider the model a black box function, h, which takes the feature vector and computes a number between 0 and 1 which represents the likelihood that Bob will reply to Alice:

$$PR_{B \rightarrow A} = h(f) \tag{5.11}$$

While AdaBoost is the model that has been shown to give the best results experimentally, this may depend on the service, and it might be worth training other types of non-parametric model such as *Random Forest* in case this is service dependent.

Finally, we use a weighted average with parameter α_A to combine $P_{A \rightarrow B}$ and $PR_{B \rightarrow A}$ into a single reciprocal score:

$$RWS_{A \leftrightarrow B}(\alpha_A) = \alpha_A \cdot P_{A \rightarrow B} + (1 - \alpha_A) \cdot PR_{B \rightarrow A} \tag{5.12}$$

The parameter α_A is defined per user, and determines which of the two scores drives successful interactions for that particular user, the user's own preferences or the probability of a reply. As an example, if Alice is a very attractive user in the general sense who rarely sends Likes, $P_{A \rightarrow B}$ is likely to be the more important score. We want to prioritise finding users who they Like, because when they do send Likes, they have a very high probability of getting replies.

We first define two sets. First, V_X is the set of users whom X has Viewed. Where $\mathcal{V}(X, Y)$ denotes a view from X to Y, then V_X is formally defined as:

$$V_X = \{Y \in U : \mathcal{V}(X, Y)\} \tag{5.13}$$

Let $\mathcal{M}(X, Y)$ be a Match between X and Y. Then, M_X is the subset of users U with whom X has Matched:

$$M_X = \{Y \in U : \mathcal{M}(X, Y)\} \tag{5.14}$$

Because it is not possible to send Likes, and therefore not possible to Match without viewing a user's profile, M_X will be a subset of V_X ($M_X \subseteq V_X$).

With these sets defined, we set our standard RRS objective of optimizing the chance of a Match by scoring the users Alice is most likely to Match with the highest. This means optimizing α_A such that the RWS score is the highest for those users.

We can turn this into a simple maximization problem based on Alice's historical data. For all the users V_A in Alice's historical data, we look to find the α_A which maximises the sum of the RWS scores of the Matches she has had with other users. By doing this, we ensure that the RWS score is as high as it can be for the average successful interaction. Formally:

$$\max_{\alpha_A} \sum_{v \in M_A} RWS_{A \leftrightarrow v}(\alpha_A) \tag{5.15}$$

In order to solve this problem, we can use one of a number of optimization algorithms. *Brent's Method* [3] is particularly suitable in this case, as it is fast within a predefined interval (in our case between 0 and 1) and does not require separate computation of derivatives. A description of these algorithms is out of the scope of this book (and they are, in any case, generally implemented in commonly used scientific computing libraries such as Python's *Scipy*), but interested readers are invited to investigate for themselves and potentially try more than one depending on how far they wish to optimize their implementation.

As a hybrid system, RWS was successful in that it showed improved performance over RCF. In particular, it was slightly less successful at generating a high volume of unidirectional preference indicators, but more successful at generating successful reciprocal preferences, which indicates that it succeeded at its original objective: to balance the preference of Alice with Bob's likelihood of replying.

5.5 Final Thoughts

We end this chapter with a short discussion of when to use hybrid systems, in particular in RRS contexts. In spite of the parallels between user-item recommender systems and RRSs, there are some additional considerations to implementing a hybrid system in this context.

Much of the literature on recommender systems indicates that hybrid systems achieve the best performance out of all types of recommender systems. Indeed, victors of recommender systems contests such as the *Netflix Prize* have often been hybrid systems. We all want the best performance out of our algorithms, so is tempting to think that constructing a hybrid system is optimal.

A recurring theme of this book is, however, that in RRS environments we suffer from a lack of public data, and a lack of research in general. As of the writing of this book, there are very few hybrid RRSs in the published literature, and the number of systems which are widely cited or appear in mainstream conferences is even fewer. For implementing an out-of-the-box hybrid system, there are therefore not very many options available to begin with.

The evaluations of these algorithms also makes it difficult to determine their advantages over vanilla content-based and collaborative filtering algorithms. All of them were evaluated on private datasets, using a variety of baselines. CCR, an early system, was evaluated using random recommendations as a baseline. That is not to say that the algorithms were not effective in the environments in which they were deployed - just that based on the existing research, it is hard to evaluate how much more effective hybrid systems are than more straightforward algorithms, especially considering the additional complexity involved in implementing and executing one.

While hybrid systems can have advantages, they also come with significant disadvantages. They are more than twice as complex to implement, as developers are required to implement two or more systems, and also a method of combining them together. They add cost to evaluation, as we need to ensure that the hybrid system is actually outperforming either of the systems which comprise it. They also cost more to run: their time complexity is bounded by the less efficient of the two algorithms, but even if all the algorithms implemented have reasonable scaling, we incur additional infrastructure costs to running multiple algorithms on the same data.

Hybrid systems are a useful tool but, especially in RRS environments, it is important not to fall into the trap of assuming they are better. Developers implementing them purely for general-case performance without first trying a simpler system are likely to be disappointed. Consider that the algorithms described in this chapter both had quite specific purposes: CCR aimed to solve the cold-start problem, and RWS aimed to balance user popularity. Hybrid RRSs are best implemented to solve a problem within an existing system, based on user feedback or performance metrics.

References

1. Akehurst, J., Koprinska, I., Yacef, K., Pizzato, L., Kay, J., Rej, T.: Ccr—a content-collaborative reciprocal recommender for online dating. In: Twenty-second international joint conference on artificial intelligence (2011)
2. Bennett, J., Lanning, S., et al.: The netflix prize. In: Proceedings of KDD cup and workshop, vol. 2007, p. 35. New York (2007)
3. Brent, R.P.: Algorithms for minimization without derivatives. Courier Corporation (2013)
4. Burke, R.: Hybrid recommender systems: Survey and experiments. User modeling and user-adapted interaction **12**, 331–370 (2002)
5. Çano, E., Morisio, M.: Hybrid recommender systems: A systematic literature review. Intelligent data analysis **21**(6), 1487–1524 (2017)
6. Doran, R.W.: The gray code. Tech. rep., Citeseer (2007)
7. Kleinerman, A., Rosenfeld, A., Ricci, F., Kraus, S.: Optimally balancing receiver and recommended users' importance in reciprocal recommender systems. In: Proceedings of the 12th ACM Conference on Recommender Systems, pp. 131–139 (2018)
8. Schapire, R.E.: Explaining adaboost. In: Empirical inference: festschrift in honor of vladimir N. Vapnik, pp. 37–52. Springer (2013)
9. Xia, P., Liu, B., Sun, Y., Chen, C.: Reciprocal recommendation system for online dating. In: Proceedings of the 2015 IEEE/ACM International Conference on Advances in Social Networks Analysis and Mining 2015, pp. 234–241 (2015)

Chapter 6
Matching Theory

6.1 Introduction

Thus far, we have largely concentrated on RRS algorithms which are based on similar structures: to find two unidirectional preferences and combine them into a single reciprocal preference. This reciprocal preference is assumed to be a number which represents the preference $P_{A \leftrightarrow B}$, which can be used to rank recommendations. While this is a robust method of making reciprocal recommendations and has been used successfully many times, there are some alternative approaches which are worth studying.

Matching Theory [5] is a field which aims to pair agents in a market where there are two sides. This has traditionally been used to solve problems where a central entity is responsible for making pairs which are generally expected to accept each other - for example, in medical residencies where students and hospitals rank each other, or organ donations where multiple criteria such as viability and geography determine who can be a recipient for which organ. It has not traditionally been applied to reciprocal recommendation, but recently a body of research has been emerging that indicates that it can be a powerful tool here as well.

The formulation of the *Stable Marriage Problem* [3] is generally where Matching Theory is considered to have initiated. It is a general problem in computer science which looks for solutions to the situation where we have n men and n women who have ranked each other in order of preference, and we want to pair everyone such that there are no two people who:

1. Are not already paired up with each other.
2. Would both rather have each other than their current partners.

Solutions to this are applicable to reciprocal recommendation. We first examine the original algorithm, designed by Gale and Shapley who first proposed the Stable Marriage Problem. We then examine a modification of this algorithm which allows it to be used in reciprocal environments

From the Stable Marriage Problem, more general forms of the problem emerged as Matching Markets, as did possible solutions to them. In Matching Markets, a resource

J. Neve, *Reciprocal Recommender Systems*, SpringerBriefs in Computer Science,
https://doi.org/10.1007/978-3-031-85103-2_6

called *Social Welfare* exists, which is contributed to by all entities in the market and in particular by matches between them. Solutions to maximise this resource can also be used to create pairs between entities in reciprocal environments.

6.2 The Stable Marriage Problem

6.2.1 Problem Description

The Stable Marriage Problem was originally formulated in 1962 [3], and is as follows. Assume we have a number of men and an equal number of women. Each man and each woman has a list of members of the opposite sex in order of preference. For example, a man would have a list of women such that the first woman on the list represented his best-case partner and the last woman represented his worst-case partner. In the problem's original formulation, every person must rank all members of their opposite sex i.e. the number of men, number of women and the length of every preference list is the same.

The objective is to pair up all men and women such that there are no men and women who are not paired, and who both prefer each other to their assigned partners.

More formally, we define two sets of size n, $X = \{x_1, x_2, \ldots, x_n\}$ and $Y = \{y_1, y_2, \ldots, y_n\}$. Then the objective is to find a set of pairs:

$$W = \{(x_i, y_j) : x_i \in X, y_j \in Y\} \tag{6.1}$$

A matching is defined by μ, which is a *bijection* between X and Y. That is to say, $\mu(x_i) = y_j$ and $\mu(y_j) = x_i$ such that μ maps x_i to one and only one y_j, and y_j to one and only one x_i.

A matching μ is said to be *stable* when there are no two values $x_a \in X$ and $y_b \in Y$ such that:

- $(x_a, y_b) \notin W$
- x_a would rather have y_b than $\mu(x_a)$
- y_b would rather have x_a than $\mu(y_b)$

The pair (x_a, y_b), where both people in the pair would rather be with each other than their current partners, is known as a *blocking pair*. In a stable matching, no blocking pairs exist. It is worth being aware that there are usually multiple possible stable matchings for any given X and Y.

The stable marriage problem is formulated as a romance-related problem. However, it is interesting to note that in the context of online dating, because we encourage users to match with as many other users as possible, the formulation of the stable marriage problem, where each user is matched with exactly one other user is slightly less useful. Instead, solutions to this problem tend to be more applicable to services such as job recommenders, where one candidate fills one vacancy, and following a successful outcome neither the candidate nor the vacancy are available any longer.

6.2.2 The Gale-Shapley Algorithm

The most well-known method for solving the stable marriage problem is the *Gale-Shapley Algorithm*, often known as the *Deferred Acceptance Algorithm* (DAA), which was outlined in the original paper that described the problem. This algorithm is optimal in the sense that it always finds a stable matching between two sets X and Y.

The algorithm's operation is as follows:

1. Initially, all members of both X and Y are *free* (i.e. no assigned matching).
2. Each free member of X, x_i proposes to the y_j which is highest on their preference list. For each proposal:

   - If y_j is free, then y_j accepts the proposal, and x_i and y_j are considered *engaged* i.e. temporarily paired.
   - If y_j is currently engaged to someone else, x_p, and y_j prefers x_p to the proposing member x_i, then the proposal is rejected and x_i remains free.
   - If y_j is currently engaged to x_p and y_j prefers x_i to x_p, then the proposal is accepted, y_j and x_i become engaged, and x_p becomes free.

3. The algorithm continues in rounds of repeating Step 2 until every x_i is engaged to a y_j.

This final set of pairings are considered the marriages, and are guaranteed to be stable. A thorough formal proof of this is outside the scope of this book, but a back-of-the-envelope proof is as follows. During the course of the algorithm, unless a set of stable pairs is found before then, every member of set X will propose to every member of set Y. Assume after running the algorithm, there exists a blocking pair (x_a, y_b). By the definition of a blocking pair, $(x_a, y_b) \notin W$. So:

1. Either x_a must at some point have proposed to y_b and been rejected for someone y_b prefers more. But then (x_a, y_b) is not a blocking pair, because y_b rejected x_a.
2. Or x_a must have been paired with someone they proposed to before proposing to y_b. But then (x_a, y_b) is not a blocking pair, because x_a proposes in order of preference.

We conclude that such a blocking pair cannot exist.

The time complexity of the algorithm is also straightforward to analyse. In the worst case, we examine the full preference lists of every x_i in X. The preference lists contain every element of Y. Both X and Y have length n, so our worst case complexity is $O(n^2)$, which is usually feasible.

While there have been alternatives to the DAA algorithm, it is considered optimal in many senses. It is guaranteed to provide stable matchings, and it does so with a single pass of the preference lists of one side of the equation. For guaranteed matchings, a better worst-case time complexity is not possible to achieve. It has some disadvantages, in particular that it is not *fair* in the sense that the individuals who propose (those in X) receive their best possible matches, whereas the individuals in Y get their worst possible partners which satisfy the conditions. While there have been algorithms which attempt to optimise for fairness, Gale-Shapley is a good starting point. Application of the stable marriage problem to reciprocal recommendation problems in any case will necessarily need some adaptation.

6.2.3 Problem Example

For the sake of clarity, we briefly present a small example problem, with three men in X, $\{x_1, x_2, x_3\}$ and three women in Y, $\{y_1, y_2, y_3\}$ looking for stable marriages.

Men	1st Choice	2nd Choice	3rd Choice
x_1	y_3	y_1	y_2
x_2	y_1	y_2	y_3
x_3	y_1	y_3	y_2

Table 6.1: Preferences for men in X

Women	1st Choice	2nd Choice	3rd Choice
y_1	x_2	x_1	x_3
y_2	x_1	x_3	x_2
y_3	x_3	x_2	x_1

Table 6.2: Preferences for women in Y

Table 6.1 and Table 6.2 show the preference lists for X and Y respectively. The objective is then to find three pairs of stable marriages. Because we have three preference lists, applying the Gale-Shapley algorithm manually we can find a set of stable pairs:

1. In the fist round:

   - x_1 proposes to y_3, and x_2 and x_3 propose to y_1.
   - y_3 accepts x_1, y_1 accepts her preferred choice x_2.
   - So we have $\{(x_1, y_3), (x_2, y_1)\}$ as engagements, with x_1 and y_2 free.

2. In the second round:

   - x_3 alone is free, and proposes to y_3 (his second choice).
   - y_3 is engaged to x_1, but prefers x_3, so she accepts him instead.
   - So we have $\{(x_2, y_1), (x_3, y_3)\}$ as engagements, with x_3 and y_2 free.

3. In the third round:

   - x_1 alone is free, and proposes to y_2 (his second choice).
   - y_2 is unmatched, and accepts the proposal.
   - So we have $\{(x_1, y_2), (x_2, y_1), (x_3, y_3)\}$ as engagements. As there are no free members, the algorithm terminates.

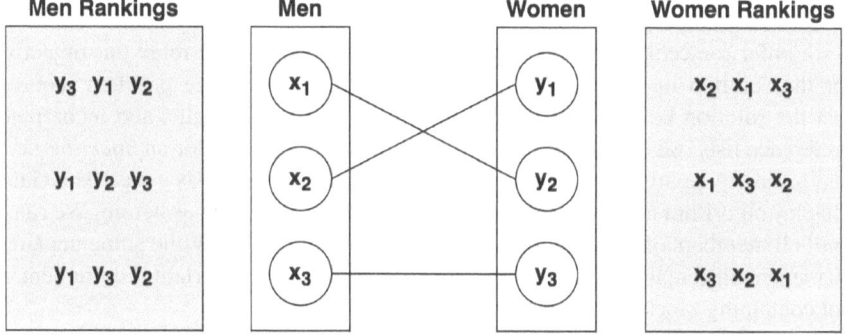

Fig. 6.1: A visualisation of the Stable Marriage Problem example

The set of engagements the algorithm terminates at, $\{(x_1, y_2), (x_2, y_1), (x_3, y_3)\}$, represents a stable matching. x_1 is the first choice of y_2, x_2 is the first choice of y_1 and x_3 is the first choice of y_3. Therefore there can be no blocking pairs. Figure 6.1 shows the example, including the preference lists and the final matching.

6.2.4 MMDAA: A Practical Application of Stable Marriage

The stable marriage problem gives us an very clear method of matching two users. However, it is not immediately evident how we might apply this to a reciprocal recommendation environment. It has been applied manually on small scales, such as to college applicants or medical students applying for residencies. In these cases, it is possible to fit the data to the problem: we can ensure that there are the same number of vacancies as applicants, and have applicants and colleges rank all of their preferences. However, it is more difficult to apply this to an online service. The data is usually sparse and unbalanced. For example, on a recruiting service the number of candidates usually far outnumbers the number of vacancies.

Multi Match Delayed Acceptance Algorithm (MMDAA) (Saini et al. [6]) is an adaptation of the DAA algorithm for reciprocal recommendation. The core of the algorithm is the original DAA algorithm. However, the input and output are modified for the purposes of an RRS environment:

- The input of the original DAA algorithm is two sets of equal sizes, X and Y, where every user in each set has ranked every user in the other set. MMDAA relaxes these constraints: the input is two sets X and Y of sizes m and n respectively, where members of each set have rankings of members of the other set that may or may not be complete.
- The output of the DAA algorithm is a list of pairs (x_i, y_j) where each item x_i in X and y_j in Y appears exactly once. The output of MMDAA is a ranked list of multiple one-to-one matches, which is more useful in the case of recommendation, where users want to see options rather than one single recommendation.

In order to accommodate the first of these two changes, we relax our objective for the solution of the problem. The original Stable Marriage problem requires that the solution be a bijection. With two sets of different lengths and incomplete preference lists, we can no longer form a bijection, so we settle for an *injection* i.e. a one-to-one mapping which does not guarantee every element has a pair. The Gale-Shapley algorithm forms this when run in exactly the same way as before. We run it until all members of every preference list has been evaluated. While some unpaired elements will remain at the end, the pairs will satisfy the important requirement of not containing blocking pairs.

The output requirement is satisfied by running the DAA algorithm repeatedly instead of just once. After generating an initial list of pairs, we consider these the highest ranked recommendations. We then eliminate the members of each pair from their respective partner's preference list, and re-run the algorithm to generate the second place recommendations (or first place for those members who were not paired up after the first round). The DAA algorithm is run k times, where k is the desired number of recommendations, or until all preference lists have been exhausted. At this point we consider our ordered list of recommendations complete and the algorithm terminates. In steps, it is as follows:

- Initialise X and Y members as free, and initial recommendation lists as empty.
- Run the modified DAA algorithm to generate a partial list of pairs of the form $\{(x_i, y_j) : x_i \in X, y_j \in Y\}$.
- For each pair (x_i, y_j), append x_i to the tail of the recommendation list of y_j, and y_j to the tail of the recommendation list of x_i.
- For each pair (x_i, y_j), eliminate x_i from the preference list of y_j, and eliminate y_j from the preference list of x_i.

- Re-initialize X and Y members as free, and re-run the DAA algorithm until the number of times run is equal to m.

This algorithm uses the DAA algorithm to generate recommendation lists based on stable matches in a system where users provide ordered lists of preferences. Where explicit preference lists are not provided by users, we can still use this algorithm to generate recommendation lists for users based on implicit preferences. In online recruiting, where this system has been successful, users generally have ways to show preference for companies and job openings - for example, by viewing the company's page, or by adding job openings to favourites lists. We can use this to generate an implicit score.

An alternative approach is to use collaborative filtering to generate the scores. If a user has a history of applying for jobs on the service, we can use a simple collaborative filtering to generate scores between 0 and 1 for other companies that the user might like. We can do the same for companies who have accepted certain applicants in the past. These collaborative filtering scores can be treated as implicit preference scores, from which we can infer preference lists between candidates and companies, with higher scores at the top of the list. The DAA algorithm can then be used to rank the recommendations in such a way as satisfies both parties.

As discussed, the DAA algorithm is $O(n^2)$ complexity. The MMDAA, which runs DAA m times in order to generate at most m recommendations is therefore $O(m \cdot n^2)$, where m is usually a relatively small number in comparison to n. The algorithm is therefore relatively efficient, and an interesting avenue of research for generating pairs in a way which does not have to incorporate traditional RRS techniques.

6.3 Matching Markets

With the introduction of the Stable Marriage Problem and its solution by Gale and Shapley, the concept of matching was consolidated from a few disjoined papers into a single, coherent interdisciplinary field. As we've seen, the algorithm itself does have direct applications into reciprocal recommendation, and has been adapted into a working RRS.

However, the formulation of the Stable Marriage Problem, which requires two sets of users, each with priority lists, is a little inflexible. Directly applying algorithms similar to the DAA algorithm to RRS environments can feel like fitting a square peg into a round hole. To do so, we must assume users provide us with clear and consistent rankings, but this is rarely the case.

The concept of *Matching Markets* is a more general and flexible extension of the concept of Stable Marriage. As in the Stable Marriage Problem, Matching Markets consider a market with two sides, where the eventual objective is to maximise the satisfaction of both sides with the matches. However, instead of preferences, we define *utility*, a resource that is transferred from one side to another upon matching, and affects the motivations of both sides to match. We also allow for the possibility that members of the market remaining unmatched could be to the advantage of the market as a whole, whereas the Stable Marriage Problem requires that everyone be matched.

Once again, we focus on recruitment and job searching services as the target for this particular class of algorithms. As with Stable Marriage, our model for Matching Markets assumes that once two users Match, their needs are satisfied and they stop looking for other potential Matches. This model, devised by Su et al. [7], fits recruitment better than online dating. In online dating, users will often aim to Match with as many other users as possible, under the assumption that only a few of these will lead to a sustained interaction, and even fewer to a relationship. The discussion is therefore framed in terms of *Candidates* and *Companies*, although this is not to say the model could not be applied to other reciprocal environments.

6.3.1 The Model

Matching markets define a *Social Welfare* resource. Social Welfare is contributed to by individual members, and we aim to maximise it across the market. This definition depends on the algorithm and the domain in which the algorithm is being applied. In general terms, the most clear social welfare function for reciprocal recommender systems is the number of Matches. Maximising the number of Matches is likely to improve the experience for the largest number of users. This section formalises this definition.

In our theoretical job market, call the set of all candidates C and the set of all employers $\mathcal{J}$. Candidates are defined as $c \in C$ and employers as $j \in \mathcal{J}$. A candidate is presented with a recommendation list of employers. The candidate looks down this list, starting with the first item, and chooses employers to apply to. Employers see the list of candidates that have applied to them, also in order of recommendation, look down the list, and pick one candidate to fill the position.

We first assume that we have used an algorithm such as those described in this book (or indeed in the user-item recommendation literature) to calculate unidirectional preference scores from candidates to employers $P_{c \rightarrow j}$ and from employers to candidates $P_{j \rightarrow c}$. For consistency with the rest of the book, we continue to use the term Like and the related formalization $\mathcal{L}(c, j)$ to express a candidate's unidirectional expression of preference towards an employer, and Match $\mathcal{M}(c, j)$ to denote that the employer has responded (i.e. both have Liked each other).

In this model, candidates are *proactive*, in the sense they always act first in applying to employers, and employers are *reactive* - they see only the candidates who have applied to them, and decide between them. A candidate decides the employer from a ranking provided, which is determined by a *ranking policy* π. The space of all possible employer rankings is denoted by $\Sigma_{|\mathcal{J}|}$, and the ranking policy is therefore:

$$\pi : C \rightarrow \Delta_{\Sigma_{|\mathcal{J}|}} \tag{6.2}$$

6.3.2 Candidates

A single ranking of employers for a candidate c under the policy π is denoted by $\sigma(c)$. The candidate receives this list, and examines employers one at a time, starting from the top of the list, until they find one that they want to apply to. We model this as an *examination function* v, which outputs the probability that an candidate c views an employer j given a particular ranking for that candidate $\sigma(c)$. v is assumed to be a convex function, the choice of which depends on the application. The probability of a candidate seeing an employer is then:

$$\mathbb{P}(\text{Candidate sees employer}) = v(\text{rank}(j|\sigma(c))) \tag{6.3}$$

We can then express the probability that a candidate Likes an employer as the probability of two subsequent events:

1. The candidate sees the employer, with probability based on the examination function $v(\text{rank}(j|\sigma(c)))$.
2. The candidate Likes the employer, with probability based on the preference score $P_{c \to j}$.

The probability of a Like occurring is therefore their product:

$$\mathbb{P}(\mathcal{L}(c, j)|\sigma(c)) = P_{c \to j} \cdot v(\text{rank}(j|\sigma(c))) \tag{6.4}$$

While current matching theory RRS methods use the preference score directly as a probability, it is worth reminding ourselves here that they are distinct concepts: the preference score is a measure calculated for ranking recommendations against each other, and does not represent the probability that a user will Like a recommendation. However, in the broader context of this definition of Social Welfare, where the probability calculation is an intermediate step to the eventual objective of finding a reasonable value to maximise, it is not unreasonable to approximate the probability as the preference score.

This gives us a probability function for a candidate Liking an employer given a specific ranking $\sigma(c)$. In order to progress towards a social welfare function which represents the whole market, we instead express this as the probability of c Liking j under particular ranking policy π. In the following equation, $\pi(\sigma(c)|c)$ represents the probability of $\sigma(c)$ appearing as a ranking for candidate c under the ranking policy π:

$$\mathbb{P}_{c,j}^{\pi} = \sum_{\sigma(c) \in \Sigma_{|\mathcal{J}|}} \pi(\sigma(c)|c) \cdot P_{c \to j} \cdot v(\text{rank}(j|\sigma(c))) \tag{6.5}$$

We then define a doubly stochastic matrix for each candidate (a matrix where each row and each column sums to 1) M_c^{π} where, the cell at (j, k) equals the probability of employer j being at position k in c's ranking list, under the ranking policy π. We can use this to further simplify Equation 6.5. We also note that the preference score $P_{c \to j}$ is independent of the ranking, so we bring it out of the sum.

$$\mathbb{P}^{\pi}_{c,j} = P_{c \to j} \cdot \sum_{k=1}^{|\mathcal{J}|} M^{\pi}_c(j,k) \cdot v(k) \tag{6.6}$$

It is useful for future computations using this formula to make an additional change for the sake of readability. First, let e_j be the standard basis in $|\mathcal{J}|$-dimensional space, with 1s in position j. Then, let $\mathbf{v}$ be the vector of size $|\mathcal{J}|$ with $v(k)$ in position k. Then, Equation 6.6 can be written as:

$$\mathbb{P}^{\pi}_{c,j} = P_{c \to j} \cdot e_j^T M^{\pi}_c \mathbf{v} \tag{6.7}$$

6.3.3 Employers

After candidates have applied to employers, employers then review applications and choose to Like (and therefore Match with) their preferred candidates. The set of candidates who have Liked an employer j under policy π is denoted by C^{π}_j. In the same way that candidates assessed employers, employer j looks down a ranking of each candidate $c \in C^{\pi}_j$ and chooses which to Match with. Unlike $\sigma(c)$, our ranking of candidates is stochastic, because we cannot be sure which candidates have applied to j.

Let $\mathrm{rank}^{\pi}_j(c)$ denote the rank of a candidate in a particular C^{π}_j. Then, as was the case with candidates choosing employers, employers take two steps to choose a candidate:

1. The employer looks down the list of candidates, and sees c with probability $\mathbb{E}[v(\mathrm{rank}^{\pi}_j(c))]$. Here we use the *expected* rank based on the probability distribution determined by the other candidates who might apply to j.
2. The employer Likes c based on their preference score $P_{j \to c}$, which we once again use to approximate the corresponding probability.

We can therefore write the probability that an employer Likes a candidate as the probability of these two events happening in sequence:

$$\mathbb{P}^{\pi}_{j,c} = P_{j \to c} \cdot \mathbb{E}[v(\mathrm{rank}^{\pi}_j(c))] \tag{6.8}$$

6.3.4 Defining Social Welfare

We can now move on to a comprehensive definition of Social Welfare (*SW*) within this system for a particular ranking policy π, as something that we can maximise. Recall that we described it informally as the number of Matches that happen in the system. A Match happens when a candidate c Likes an employer j, and j subsequently Likes c. Under the ranking policy π, this happens with probability

$\mathbb{P}^{\pi}_{c,j}\mathbb{P}^{\pi}_{j,c}$. Over all candidates and employers, the expected number of Matches is therefore:

$$SW(\pi) = \sum_{c \in C} \sum_{j \in \mathcal{J}} \mathbb{P}^{\pi}_{c,j} \mathbb{P}^{\pi}_{j,c} \tag{6.9}$$

Substituting Equation 6.7 and Equation 6.8, we can write the social welfare function as:

$$SW(\pi) = \sum_{c \in C} \sum_{j \in \mathcal{J}} (P_{c \to j} \cdot e_j^T M_c^{\pi} \mathbf{v})(P_{j \to c} \cdot \mathbb{E}[v(\text{rank}_j^{\pi}(c))]) \tag{6.10}$$

Equation 6.10 is quite long, but most of the individual terms are not particularly challenging to compute. The exception is $\text{rank}_j^{\pi}(c)$, because it depends on the other candidates applying to employer j. The following paragraphs explain how to approximate this term.

We define a ranking function φ such that $\varphi_j(c)$ is the overall rank of candidate c for employer j (i.e. if all candidates applied to j, c would be in position $\varphi_j(c)$). The inverse function $\varphi_j^{-1}(s)$ is the candidate at rank s for employer j under the same conditions i.e. $\varphi_j^{-1}(s) = \{c' \in C : \varphi(c') = s\}$. The probability of a candidate c having rank k is therefore dependent on the probabilities of each $c' \in C$ applying to j, where $\varphi(c') < \varphi(c)$.

With this definition in mind, we can model the rank of c as viewed by j under π as $\text{rank}_j^{\pi}(c) \sim 1 + X_{j,c}^{\pi}$ where $X_{j,c}^{\pi}$ is a Poisson Binomial random variable with parameters $[\mathbb{P}_{\varphi_j^{-1}(1)}, \mathbb{P}_{\varphi_j^{-1}(2)}, \cdots, \mathbb{P}_{\varphi_j^{-1}(\varphi_j(c)-1),j}]$. The notation here becomes a little dense, but the meaning is more straightforward: $\mathbb{P}_{\varphi_j^{-1}(1)}$ is the probability that the candidate most preferred by j applies to j. Our list of parameters is therefore each of the probabilities that a candidate c' where $\varphi(c') < \varphi(c)$ applies to j.

For a formula for $\mathbb{P}(\text{rank}_j^{\pi}(c) = k)$ we define two more terms. First, $A_j(c) = \{c' \in C : \varphi(c') < \varphi(c)\}$ is the set of candidates who have ranks that are higher in j's overall ranking than c. Then $F_{(j,c)}^l = \{B \subseteq A_j(c) : |B| = l, l \leq |A_j(c)|\}$ is the set of all subsets of of l items that can be selected from $A_j(c)$.

In order for a candidate c to end up in position k in the list of employer j, we need exactly $k - 1$ candidates from $A_j(c)$ to apply to j. If more than that apply, c ends up at a higher rank, and if fewer than that apply, c will be at a lower rank. For a given set $U \subseteq A_j(c)$ of size $k - 1$, c will be at position k if all the elements of U applied to j, and all the elements of $A_j(c) \notin U$ didn't apply to k:

$$\mathbb{P}(\text{rank}_j^{\pi}(c) = k | \mathcal{L}(c', j) \forall c' \in U) = \prod_{s \in U} P^{\pi}_{s \to j} \prod_{r \in A_j(c) \setminus U} (1 - P^{\pi}_{r \to j}) \tag{6.11}$$

The probability that c has rank k is therefore the sum over all possible subsets of $A_j(c)$ of length $(k - 1)$:

$$\mathbb{P}(\text{rank}_j^\pi(c) = k) = \sum_{U \in F_{(j,c)}^{k-1}} \prod_{s \in U} P_{s \to j}^\pi \prod_{r \in A_j(c) \setminus U} (1 - P_{r \to j}^\pi) \qquad (6.12)$$

While this computation is possible, calculating it for every member of every subset involves $\frac{|A_j(c)|!}{(|A_j(c)|-l)!l!}$ terms, which quickly becomes intractable, especially if we end up computing it over a large number of candidate-employer pairs. Instead, a lower bound is generally used in the social welfare equation:

$$SW(\pi) = \sum_{c \in C} \sum_{j \in \mathcal{J}} (P_{c \to j} \cdot e_j^T M_c^\pi \mathbf{v}) \cdot$$
$$(P_{j \to c} \cdot v(1 + \sum_{c' \in A_j(c)} P_{c' \to j} \cdot e_j^T M_{c'}^\pi \mathbf{v})) \qquad (6.13)$$

The proof of this lower bound is omitted from this book for length reasons, but interested readers can see it described by Su et al. in the paper which introduced this method.

6.3.5 Maximising Social Welfare

Our final challenge is to maximise Equation 6.13. Recall that maximising the Social Welfare equation is equivalent to maximising the number of Matches in the system. We present a straightforward method for doing so here, and end by briefly discussing a more efficient approximation, with the caveat that as of the time of writing, this is an active research area, and new methods are being developed every year.

A straightforward way to maximise social welfare is to use gradient descent. This is similar to the way that latent factor models were computed in Section 3.4.2. The slight difference is that we are not aiming to minimize error, but to maximise Social Welfare. We therefore take small steps in the direction of the positive gradient, which should increase SW with each iteration.

We start by initializing M_c^π to a doubly stochastic matrix such as the uniform matrix. In order to maximise Social Welfare, first step is to calculate the the partial differential of $SW(\pi)$ with respect to a specific $M_c^\pi(j,k)$, which will give us the gradient:

$$\frac{\partial SW(\pi)}{\partial M_c^\pi(j,k)} = P_{c \to j} \cdot P_{j \to c} \cdot v(1 + \sum_{c' \in A_j(c)} P_{c' \to j} e_j^T M_{c'}^\pi \mathbf{v}) \qquad (6.14)$$

Having computed the gradient, we can then use this to take a step in the direction which increases social welfare. Using a learning rate η which determines the size of the step taken, the update function is:

$$M_c^\pi = M_c^\pi + \eta \Delta SW(\pi) \qquad (6.15)$$

This update rule increases the social welfare along the gradient, but does not preserve the doubly stochastic property of the matrix. Between update steps, we therefore project the updated matrix back onto the *Birkhoff Polytope* (the set of all doubly stochastic matrices) [1]. A simple method of doing this is the *Sinkhorn-Knopp Algorithm* [4], which alternately normalizes rows and columns by dividing each value in a row or column by that row or column's sum, converging on a matrix where both rows and columns sum to 1.

For example, given an arbitrary 3×3 matrix:

$$\begin{pmatrix} 1 & 2 & 3 \\ 4 & 5 & 6 \\ 7 & 8 & 9 \end{pmatrix} \tag{6.16}$$

We normalize the rows by dividing them by their sum:

$$\begin{pmatrix} \frac{1}{6} & \frac{2}{6} & \frac{3}{6} \\ \frac{4}{15} & \frac{5}{15} & \frac{6}{15} \\ \frac{7}{24} & \frac{8}{24} & \frac{9}{24} \end{pmatrix} \tag{6.17}$$

We then repeat this process for the columns, dividing each element of each column by the sum of the respective column. The Sinkhorn-Knopp Algorithm is guaranteed to converge on a doubly stochastic matrix in polynomial time, which represents a projection of the original matrix onto the Birkhoff Polytope.

The three steps of computing the gradient, taking a step of the size of the learning rate in the positive direction to maximise the $SW(\pi)$ function and then projecting the resulting matrix to a doubly stochastic matrix are continued until M_c^{π} converges.

There are a number of alternative methods of doing this optimization. Su et al. also present positive results using the *Frank-Wolfe Algorithm* [2]. More recently, indirect methods have also found success. In particular, Tomita et al. discuss an effective model using *Transferable Utility* to maximise social welfare [8, 9]. This method is interesting because it allows us to efficiently model not only the importance of preferences, but also the importance of the transfer of money during the hiring process, an important factor which may influence a candidate to take a job in spite of a lack of preference.

This model assumes that a candidate has a preference for a company $P_{c \to j}$ and vice-versa $P_{j \to c}$. The model also assumes a level of error in each of these preferences $E_{c \to j}$ and $E_{j \to c}$, which represents both the error in our preference estimation, and the noise in the model in general (candidates and companies often do not have preferences which they stick to rigidly, and may be swayed by factors such as their mood on a particular day) which is modeled by a random distribution. Finally, there is a transfer of resources between the candidate and the company $T_{c,j}$ which happens upon hiring, and represents money as well as benefits etc. which the company has to pay for. When a candidate and a company match, the candidate obtains a *Utility* of:

$$P_{c \to j} + E_{c \to j} + T_{c,j} \tag{6.18}$$

And the employer obtains a Utility of:

$$P_{j \to c} + E_{j \to c} - T_{c,j} \tag{6.19}$$

The $T_{c,j}$ on the employer's side is because they transfer money to the candidate during the hiring process, so they receive the value they get from the candidate, minus the value of the salary they pay. From this, we can derive a system of equations that we can then optimize to find an equilibrium where candidates and employers all obtain a satisfactory Utility.

6.4 Final Thoughts

It is easy to imagine that these kinds of models which apply matching theory to reciprocal recommendation are flexible and powerful: we can modify the Social Welfare function and the optimization method to fit our particular service, depending on what we want to maximise and what we hope users themselves want to get out of using the service.

From a research point of view, Matching Theory models are also in many ways more interesting and exciting than those described in other parts of this book. It represents a deviation from the standard model of simply adapting user-item recommender systems methods to the reciprocal domain. It also makes use of academic theory which was originally conceived with matching people in mind, before the existence of social recommender systems or even the Internet.

However, the jury is still out on the effectiveness of Matching Theory algorithms as opposed to more proven techniques such as collaborative filtering. While research does indicate that these methods are effective, the experimental results are not totally convincing. Many of the evaluations of these methods have been based on offline testing on synthetically generated data, which is not always a good predictor of a real user's response to recommendations when presented with them. Where offline testing on real data has been conducted, it has been on datasets of a few hundred users at most.

As with hybrid systems, this is not to say that the models described in this chapter are not effective, just that more research is required to determine whether their implementation is worth the additional complexity over and above something like vanilla collaborative filtering. It is likely that there are cases where it is worth it, but this is something that should be investigated thoroughly by testing on a service-to-service basis.

References

1. Birkhoff, G.: Tres observaciones sobre el algebra lineal. Univ. Nac. Tucuman, Ser. A **5**, 147–154 (1946)

2. Frank, M., Wolfe, P., et al.: An algorithm for quadratic programming. Naval research logistics quarterly **3**(1-2), 95–110 (1956)
3. Gale, D., Shapley, L.S.: College admissions and the stability of marriage. The American Mathematical Monthly **69**(1), 9–15 (1962)
4. Knight, P.A.: The sinkhorn–knopp algorithm: convergence and applications. SIAM Journal on Matrix Analysis and Applications **30**(1), 261–275 (2008)
5. Lovász, L., Plummer, M.D.: Matching theory, vol. 367. American Mathematical Soc. (2009)
6. Saini, A., Rusu, F., Johnston, A.: Privatejobmatch: a privacy-oriented deferred multi-match recommender system for stable employment. In: Proceedings of the 13th ACM Conference on Recommender Systems, pp. 87–95 (2019)
7. Su, Y., Bayoumi, M., Joachims, T.: Optimizing rankings for recommendation in matching markets. In: Proceedings of the ACM Web Conference 2022, pp. 328–338 (2022)
8. Tomita, Y., Togashi, R., Hashizume, Y., Ohsaka, N.: Fast and examination-agnostic reciprocal recommendation in matching markets. In: Proceedings of the 17th ACM Conference on Recommender Systems, pp. 12–23 (2023)
9. Tomita, Y., Yokoyama, T.: Fair reciprocal recommendation in matching markets. In: Proceedings of the 18th ACM Conference on Recommender Systems, pp. 209–218 (2024)

Chapter 7
Ethical Concerns and Future Work

7.1 Introduction

This book has explored a number of methods of designing reciprocal recommender systems. We have explored content-based, collaborative filtering and hybrid systems. We have also had a look at some Matching Theory approaches to reciprocal recommendation, and how algorithms from other fields such as economics have been applied to RRSs.

We end this book with a short chapter focusing on two discussions. Firstly, we consider the ethics of reciprocal recommendation. Person-to-person matching has some considerations which are not always present in user-item recommendation, and there are sparse mentions in the literature of the ethical concerns relating to reciprocal recommendation.

Finally, we look at some directions for future work in reciprocal recommendation. While there is an increasing body of research on reciprocal recommender systems, the field is still relatively sparse compared to the body of research on user-item recommender systems. This section discusses some of the areas which are currently lacking, and which therefore present opportunities for people working in this field.

We first discuss areas in content-based and collaborative filtering which are lacking in research. There are some particularly important gaps in the current literature with regard to bringing existing user-item recommenders to reciprocal environments, which could be straightforward to fill by someone with access to data and the resources to design and test RRSs.

There are also some other topics outside of the effectiveness of specific systems which have not been sufficiently explored by existing research. These include both problems which also exist in user-item recommender systems such as the *Cold-Start Problem* [12] and *Serendipity* [7] and problems that are unique to reciprocal environments, such as accounting for user active and passiveness.

Finally, there are some global issues with the current landscape of RRS research. The most important of these is the lack of general datasets on which to test multiple algorithms to confirm their effectiveness with respect to each other, such as exist

J. Neve, *Reciprocal Recommender Systems*, SpringerBriefs in Computer Science,
https://doi.org/10.1007/978-3-031-85103-2_7

in user-item recommender systems. Connected to this is how reproducible results are: the same algorithms implemented in different papers often give quite different results.

7.2 Ethical Considerations in Reciprocal Recommendation

Reciprocal recommendation is ethically a particularly sensitive topic. The primary data used is identifiable user data, which leads to privacy and security concerns. In addition, the targets are also users, which can lead to concerns about bias.

There is currently no published research regarding the ethical challenges of specifically reciprocal recommender systems. However, there has been research on user-item recommender systems ethics, which has identified a number of concerns. Many of these are amplified in reciprocal environments because of the sensitivity of person-to-person relationships. It is important that algorithm designers are aware of these issues and attempt to mitigate them as far as possible in their data preparation and system design.

7.2.1 Privacy

One of the main concerns when working with user data is privacy and security of that data. This is particularly difficult when working with content-based filtering in reciprocal recommendation environments. In many cases where working with user data is required, it is possible to anonymise this data so that if outside parties somehow gain access to it, they are not able to identify specific users. This is often not possible in content-based filtering RRS environments, because information that could potentially identify users is a necessary part of the dataset. Significant precautions must therefore be taken to ensure that the data is seen by the smallest number of people, and the users are aware that identifying data will be used as part of the recommender system.

7.2.2 Fairness and Bias

Fairness and bias is a common problem in recommender systems, where some items or groups of items are disproportionately popular, and therefore repeatedly recommended, and other items are rarely recommended. It can also be an issue if certain user groups are over- or under-represented in the data [1]. This is most common in collaborative filtering, but also occurs in content-based filtering, where certain attributes may make certain users or types of users more popular than others.

This becomes a problem especially when certain categories of users, for example, users of a certain race, are disproportionately recommended by the system. If certain types of users rarely appear in recommendations, this can reinforce negative stereotypes about these users. It can also discourage them from using the service if, as a result of not appearing in recommendations, they receive very little attention. Finally, users often notice this kind of bias themselves from looking at their recommendations every day.

7.2.3 Appearance-Based Recommendation

There has been very little research in general on machine learning for popularity or personal attractiveness in the context of photos. However, there may be ethical concerns where basing recommendations on appearance is concerned. While some online dating services encourage users to make decisions based on appearance, others aim to promote connections based on compatibility of personality. Basing recommendations on appearance could potentially reinforce attraction based on superficial physical features. This might increase the number of matches, but be incompatible with the ultimate goals of users of finding a significant long-term relationship.

In addition, the most physically attractive users of both genders tend to command disproportionate attention in online dating services, with a high percentage of Likes often going to a small percentage of users with appealing photos. Although score aggregation using operators such as the Harmonic Mean which penalise large differences in scores should combat this to some extent [9], it is also easy to imagine that if Bob has a disproportionate number of high scores where unidirectional preference ($P_{x \to \text{Bob}}$) is concerned, he has a higher chance of appearing on a lot of users' recommendation lists.

7.3 Content-Based Filtering

As discussed in Chapter 4, there have been a number of algorithms developed to perform reciprocal recommendation using content-based filtering methods. The two case studies we examined were *RECON* [11], which makes recommendations based on categorical data, and *ImRec* [8], which makes recommendations based on images.

7.3.1 Categorical Data

Where categorical data is concerned, there is a wide range of potential improvements to be exploited. We start with RECON, which was a very early RRS, and which is often used as a baseline for other algorithms. Even so, there are a number of areas where relatively simple improvements have not yet been tested. Two of them are outlined below:

1. Where continuous data such as user age appears in the profile, RECON buckets the data to turn it into categorical data (e.g. (20-24), (25-29) and so on). However, this treats all categories outside of the one which matches the user under consideration as the same. For example, if Bob is 25 years old, it is likely that Alice, who has so far exclusively Liked users in the 20-24 category, might still find him appealing. However, under RECON, his preference score will be penalised to the same degree as someone who is 60 years old. We could improve this by defining a distance metric such that users further away in age from a user's preferred range receive a lower preference score.

2. The number of possible values an attribute could take is likely to have an impact on how significant attribute matches are, as are the proportions of people on the service who possess those attributes. If there are two possibilities for 'Smoker', 'Yes' and 'No', and Alice has liked 5 people who are non-smokers, this may be a coincidence, especially if most people on the service are non-smokers. However, if the service has 30 possible professions to choose from, and Alice has Liked 5 people who have their profession listed as 'Musician', an unusual full-time occupation, it is likely that this is an important quality for her. Nonetheless, a vanilla implementation of RECON would treat these two the same. We could improve this by adjusting scores based on the number of possible options, as well as how uncommon it is to find users who have chosen those options.

A number of papers use improvement on RECON as a baseline for showing a new algorithm's success. However, we still do not know whether such algorithms would outperform a variant of RECON where such low-hanging fruit optimizations as those above were made.

In addition to optimizing algorithms such as RECON, which treat attributes as independent variables all of which hold equal predictive power, there is also scope for more research on categorical data-based recommender systems based on machine learning models such as Neural Networks. When trained on large datasets, these tend to be more adept at capturing the importance of different attributes, as well as potential relationships between attributes (e.g. if Alice mostly Likes people with pets, she probably likes animals and therefore might be interested in someone whose profession is 'Vet').

7.3.2 Unstructured Data

There have been several papers investigating the use of user images as a predictor of attraction for the purposes of recommendation. However, there is still potential scope for more work on this. In addition to personal attraction, based on the image directly, there is the possibility of extracting individual attributes from images such as facial expression, lighting, image quality and so on. These could potentially be used both for recommendation, and to help users improve their own profiles and attract more positive attention.

There is very little research that has been done on the use of freetext profiles in reciprocal recommendation. This is a shame, as it is quite likely to be a strong predictor of indicators of interest. This is especially true in environments outside of online dating such as recruitment, where photos are much less important. Freetext sections such as company job descriptions, and descriptions of a user's responsibilities at previous jobs are quite likely to be as strong a predictor of interest from a company as categorical data such as qualifications. Using a transformer-based model such as BERT [2] to extract embeddings and use them to make predictions about reciprocal preferences could potentially be a successful method of reciprocal recommendation which is yet to be explored.

7.4 Collaborative Filtering

Collaborative filtering methods in user-item recommender systems are quite advanced. Each year significant effort is put into developing new methods of collaborative filtering on large datasets such as the *MovieLens* [6] dataset in order to improve on previous performance metrics by fractions of a percent, to the point where there is some doubt about whether users would even notice these fractional improvements.

In comparison, the landscape of reciprocal recommender systems is much sparser. There are two important facets to this: the data available, and the algorithms themselves. A brief discussion of both is provided here.

7.4.1 Data and Reproducibility

Where user-item recommender systems are concerned, there are a number of large, public datasets which are considered the standard on which to test new systems. Particularly common are the *Netflix* dataset and the *MovieLens* dataset. Both of these contain 100 million movie ratings (with smaller versions also available), and the latter also contains metadata such as genres and tags. These datasets are widely considered a benchmark for new recommender systems and, while algorithms developed for specific niches may use other datasets, being able to reproduce and test a published algorithm with the same parameters on these datasets and get a very similar result is useful for the advancement of the field in general. (It is worth noting, however, that even when algorithms are re-implemented on these datasets, they do not always yield the same results as they did in their original papers [3].)

Privacy concerns around person-to-person data prohibit this from being possible in reciprocal environments. It is not possible to remove identifying information from online dating or online recruitment while still retaining all the salient information in the same way that it is for movie ratings. The data is, in any case, much more sensitive, with privacy being a key concern of users of person-to-person services where RRSs might be used.

This means that there are no public datasets on which users can compare multiple RRSs against each other. This is a particularly significant disadvantage for collaborative filtering algorithms. While the performance of content-based algorithms is to some extent expected to be dependent on the system in which they are deployed, the performance of collaborative filtering algorithms is less likely to change significantly relative to other algorithms when they are deployed on different datasets. It is therefore a significant disadvantage that there is no single public dataset on which they can all be implemented and compared.

This issue is compounded by the fact that RRS algorithms have a tendency to perform significantly better in the papers in which they were originally described, compared to when they are subsequently implemented as points of comparison for other algorithms. This phenomenon is not unique to RRS, and does not imply that the implementations were incorrect or the results dubious. For example, RECON has been quite effective on some services, but it might underperform on a service where users make decisions based on photos without ever seeing categorical data. Researchers often tinker with an algorithm until they achieve a good result on their particular dataset, and sometimes fine tune hyperparameters of baseline algorithms with less enthusiasm. However, it is unfortunate that details of these datasets are never published. Because of this, it is impossible to effectively compare algorithms with each other: if I design an algorithm which outperforms the state of the art on my dataset, we have no way of knowing if it is more effective in general, or just on that particular dataset. If the latter, we cannot analyse what characteristics of the dataset resulted in the difference in performance. The result is a number of algorithms with no clear best choice - just a series of fluctuating performances on different datasets.

Because we have no public datasets with which to make direct comparisons between algorithms, future research on reciprocal recommendation would benefit from more details about the specifics of the dataset being listed, to the extent that this is possible within the bounds of reasonable concern about privacy. In addition, where trained models are used, specifics on essential training information such as hyperparameters is often omitted from papers. Going forward, this information being more rigorously detailed in research works would help researchers who wish to compare performance metrics of algorithms.

7.4.2 Algorithm Design

As discussed above, the field of collaborative filtering algorithms in user-item recommender systems is relatively advanced, and a wide range of technologies have been adopted to solve the problem of recommending items to users in different contexts. A thorough treatment of every algorithm which has been successful at user-item recommendations which has never been applied in reciprocal environments would be too long for this book, so we discuss a couple of areas where further research is needed.

Where nearest neighbour algorithms are concerned, there are a number of optimizations that have not been tested in reciprocal environments. In particular, in Chapter 3, we discussed the algorithm RCF [15], and how its scaling made it challenging to use in larger services. Where nearest neighbour algorithms are used in services with many users, techniques such as clustering users and applying the algorithm within clusters are often used to combat these inefficiencies. For example, on dating services, users tend to date other users within their geographical area, so the effect of running a nearest neighbour algorithm on a service where users were clustered by area would be interesting to see.

Where latent factor models are concerned, the method of stochastic gradient descent for constructing latent factor matrices from a preference matrix is now considered a little outdated. More recent approaches utilise modern machine learning methodologies to improve on the accuracy of this, especially on sparse matrices. There are numerous examples of this, but a representative and often cited one is *DeepFM* [5], which uses a combination of a factorization machine and deep neural network in parallel to make predictions. This is often much more effective at capturing both pairwise and higher order interactions between users for the purposes of making predictions.

7.5 Hybrid Systems

Hybrid systems still have quite a long way to go as far as coverage in reciprocal environments goes. There are very few examples of hybrid systems in the literature that are convincingly better than their vanilla counterparts. To the best of the author's knowledge, there are no examples in the literature of effective switching hybrid or mixed hybrid algorithms.

Future research could explore algorithms similar to RWS - those that use hybridization to solve a specific problem such as popularity within the context of online dating. Similar issues such as *Serendipity* [4] (showing users recommendations which they would not usually search for) and *Passiveness* [10] (users who will reply to others but who will rarely initiate Likes themselves) exist in reciprocal environments, and it would be interesting to see hybrid systems which directly improved on the results of the systems which comprise them in some specific way, rather than just showing improved performance metrics against an arbitrary baseline.

7.6 Matching Theory

Matching theory-based approaches to reciprocal recommendation are a relatively recent area of research and, as such, there is significant ground to cover here. A few points in particular stand out here. Firstly, more rigorous evaluation is required to establish the performance of matching theory-based methods in reciprocal environments as compared to alternative methods. Much of the research is based on automatically generated synthetic data, which is rarely representative of performance when a system is used to present recommendations to real users. Where tests on offline data have been performed, it is usually not more than a few hundred users. An evaluation of the effectiveness and efficiency of the algorithms developed so far over existing approaches on a more comprehensive dataset would make matching theory approaches significantly more convincing.

Most of the current approaches to designing RRSs using matching theory have focused on employers and candidates, in the standard recruitment model where candidates act, followed by employers [13, 14]. This is often not directly applicable to other RRS services such as online dating, and Social Welfare functions could be designed for online dating and other RRS environments. This might lead to different approximation methods being effective. For example, Transferable Utility might not be as effective in an online dating environment, as no money is transferred directly, or conversely it might be an effective method of modelling the effect of popularity on a user's likelihood of responding to an indicator of preference.

References

1. Chen, J., Dong, H., Wang, X., Feng, F., Wang, M., He, X.: Bias and debias in recommender system: A survey and future directions. ACM Transactions on Information Systems **41**(3), 1–39 (2023)
2. Devlin, J.: Bert: Pre-training of deep bidirectional transformers for language understanding. arXiv preprint arXiv:1810.04805 (2018)
3. Ferrari Dacrema, M., Cremonesi, P., Jannach, D.: Are we really making much progress? a worrying analysis of recent neural recommendation approaches. In: Proceedings of the 13th ACM conference on recommender systems, pp. 101–109 (2019)
4. Ge, M., Delgado-Battenfeld, C., Jannach, D.: Beyond accuracy: evaluating recommender systems by coverage and serendipity. In: Proceedings of the fourth ACM conference on Recommender systems, pp. 257–260 (2010)
5. Guo, H., Tang, R., Ye, Y., Li, Z., He, X.: Deepfm: a factorization-machine based neural network for ctr prediction. arXiv preprint arXiv:1703.04247 (2017)
6. Harper, F.M., Konstan, J.A.: The movielens datasets: History and context. Acm transactions on interactive intelligent systems (tiis) **5**(4), 1–19 (2015)
7. Kotkov, D., Wang, S., Veijalainen, J.: A survey of serendipity in recommender systems. Knowledge-Based Systems **111**, 180–192 (2016)
8. Neve, J., McConville, R.: Imrec: Learning reciprocal preferences using images. In: Proceedings of the 14th ACM Conference on Recommender Systems, pp. 170–179 (2020)
9. Neve, J., Palomares, I.: Aggregation strategies in user-to-user reciprocal recommender systems. In: 2019 IEEE International Conference on Systems, Man and Cybernetics (SMC), pp. 4031–4036. IEEE (2019)
10. Pizzato, L., Rej, T., Akehurst, J., Koprinska, I., Yacef, K., Kay, J.: Recommending people to people: the nature of reciprocal recommenders with a case study in online dating. User Modeling and User-Adapted Interaction **23**, 447–488 (2013)

11. Pizzato, L., Rej, T., Chung, T., Koprinska, I., Kay, J.: Recon: a reciprocal recommender for online dating. In: Proceedings of the fourth ACM conference on Recommender systems, pp. 207–214 (2010)
12. Schein, A.I., Popescul, A., Ungar, L.H., Pennock, D.M.: Methods and metrics for cold-start recommendations. In: Proceedings of the 25th annual international ACM SIGIR conference on Research and development in information retrieval, pp. 253–260 (2002)
13. Su, Y., Bayoumi, M., Joachims, T.: Optimizing rankings for recommendation in matching markets. In: Proceedings of the ACM Web Conference 2022, pp. 328–338 (2022)
14. Tomita, Y., Togashi, R., Hashizume, Y., Ohsaka, N.: Fast and examination-agnostic reciprocal recommendation in matching markets. In: Proceedings of the 17th ACM Conference on Recommender Systems, pp. 12–23 (2023)
15. Xia, P., Liu, B., Sun, Y., Chen, C.: Reciprocal recommendation system for online dating. In: Proceedings of the 2015 IEEE/ACM International Conference on Advances in Social Networks Analysis and Mining 2015, pp. 234–241 (2015)